CORRIDORS OF TIME

From Alligator Butte

Early morning in the heart of Grand Canyon. About 3,000 feet above the Colorado River, and 2,000 feet below Hopi Point on the South Rim.
1. E, F-4.

READER'S DIGEST EDITION

CORRIDORS OF TIME

1,700,000,000 YEARS OF EARTH AT GRAND CANYON

PANORAMIC PHOTOGRAPHY AND TEXT BY

RON REDFERN

ILLUSTRATIONS BY

GARY HINCKS

INTRODUCTION BY

CARL SAGAN

Times BOOKS

This book was made possible by many generous Americans who gave the author unstinting help, advice, encouragement, enthusiasm, and warm hospitality. It is to these marvelous people that this volume is dedicated.

Illustrations pp. 86–87 from Fossil Behavior *by Adolf Seilacher:*
Copyright © August, 1967 by Scientific American, Inc.,
all rights reserved; pp. 118–119 from The Oldest Fossils *by Elso S. Barghoorn:*
Copyright © May, 1971 by Scientific American, Inc.,
all rights reserved; pp. 120–121 from Crises in the History of Life *by Norman D. Newell:*
Copyright © February, 1963 by Scientific American, Inc., all rights reserved.

Illustrations pp. 166, 166–167 (bottom), 168–169 (bottom and top left):
Courtesy of Beinecke Rare Book Library, Yale University;
p. 167: Courtesy of Jefferson Expansion Memorial, St. Louis, National Park Service;
p. 169 (top): Courtesy of the U.S. Geological Survey

Printed by The Reader's Digest Association, Inc.
Originally published by Times Books

Library of Congress Catalog Card Number 82-62208
ISBN 0-89577-167-5

Printed in the United States of America

CONTENTS

INTRODUCTION BY CARL SAGAN		6
PREFACE		7
I	THE NEW PANGENESIS	10
II	CORRIDORS OF TIME	30
III	1,700,000,000 YEARS OF EARTH	50
IV	THE LOST MILLENNIA	64
V	OF TRILOBITES AND BRACHIOPODS AND GREAT RED WALLS	82
VI	SON OF JUPITER	96
VII	SPECTRUM OF LIFE	112
VIII	THE MAGIC NUMBER	132
IX	INDIGENOUS MAN	150
X	INTRUSIVE MAN	164
ACKNOWLEDGMENTS		194
BIBLIOGRAPHY		194
INDEX		196
MAPS	THE COLORADO PLATEAU	14,19

INTRODUCTION

CARL SAGAN

Everything is connected with everything else. If we look deeply enough, we can find strands of causality binding together any two objects or events in the Cosmos. But occasionally we come upon a thing which speaks to us eloquently of profound connections, both in time and space. The Grand Canyon of northwestern Arizona was cut over the ages by the Colorado River. The exposed rocks in the Canyon walls have a story to tell, if we are willing to pay attention; and clarify the processes which create and destroy the landscapes of Earth. There are Precambrian formations from the time before there were any individual plants or animals big enough to see; there are Mesozoic rocks, from the time when hissing reptiles lumbered through the steaming jungles; there is Sunset Crater, a volcano which erupted in the eleventh century A.D., burying in ash a village made by creatures who evolved on Earth only a few million years earlier. Much of the history of our planet can be found in the Grand Canyon section of the Colorado Plateau.

Ron Redfern is an amateur in the best sense of the word. For more than a decade he has been pursuing and nurturing a love for this exquisite geological feature—reading about it, exploring it himself, talking to the old hands and scientific experts in the lore of the Grand Canyon. In addition, he has developed a breathtaking sequence of panoramic photographs, approaching a full 360 degrees, illuminating the stratigraphic and other wonders of the Canyon. Redfern has sought profound connections in the Grand Canyon region. He does not discuss only the superficial geological processes which have produced what we see there, but also connects it with questions of continental drift, plate tectonics, and mantle convection, the grand motions in the interior of the Earth which produced as minor consequences some of the geological features, such as North America, which we find impressive, or at least familiar. Not only does he describe the diversity of contemporary life forms to be found in the Grand Canyon today; he traces the fossil record in the Canyon back more than a billion years to colonial blue-green algae; and then carries us three billion years earlier still, to the origin of life on Earth. Not only do we learn of John Wesley Powell's scientific exploration of the Grand Canyon in 1869, and the investigations by his scientific successors, we are also told of the earliest traces of human habitation in the vicinity of the Canyon.

And forward in time as well: Redfern takes us to the Apollo astronauts who trained for their traverses of the lunar surface by wandering through the exquisite geology of the Grand Canyon. Repeatedly we are reminded that similar processes are working on other worlds as they are here on Earth. The astonishing sulphur vulcanism of Io, the innermost large moon of Jupiter, is tied to the volcanic processes of Arizona. Redfern imagines a Viking Lander equipped to search for life on Mars settling down in the Grand Canyon and discovering the spectacular microbiology of Earth. But for all its beauty and grandeur the Grand Canyon is only 350 kilometers long. There seems to be a much longer canyon on Venus. And on Mars there is a feature called *Vallis Marineris*—Mariner Valley—discovered by the Mariner 9 spacecraft, the first vehicle to orbit another planet. Vallis Marineris, if emplaced on the Earth, would stretch from New York to San Francisco, a distance of 5,000 kilometers, and the Grand Canyon of Arizona would then fit into one of its minor tributary valleys. Perhaps one day there will be a book on Vallis Marineris which connects it to the history and geology of Mars and exhibits lovely panoramic photographs as Ron Redfern has done for the more modest Grand Canyon which graces our world.

PREFACE

RON REDFERN

In 1754 Horace Walpole wrote a fairy story called *The Three Princes of Serendip*, whose heroes "were always making discoveries, by accidents and sagacity, of things they were not in quest of," a knack that came to be called "serendipity." In many ways this book is an example of that art. All my life I have been blessed with a feeling of inadequacy, the result of a brief period of blindness in late childhood as well as a half-lifetime of deafness. I was fortunate enough to recover both my sight and almost normal hearing, but not before I had learned a lesson or two.

The advantage of feeling inadequate is that one has nothing to lose by trying to achieve the impossible. The art of serendipity depends on not knowing when to stop. The effect of their interaction can be surprising; photography is a good example of the way in which the formula works. About twenty-five years ago I was asked to explain in a magazine feature how something was made. I thought it wise to assist my inadequate words with a photograph or two, but I had never before taken a serious photograph. Nevertheless, I tried and, to my surprise, I eventually found myself writing and illustrating a series of articles and making a few film documentaries besides.

After a ten-year interval, in which photography did not feature in my work at all, I visited Grand Canyon, a brief stop in a wild fandango of a seven-day tour of the "If-it's-Tuesday-it-must-be-Brussels" variety. My business in England at that time necessitated frequent visits to the United States. What I didn't realize was that to return to Grand Canyon for a more relaxed look would be unwise. If there is one place on Earth that is likely to increase one's feeling of inadequacy it is the Grand Canyon of the Colorado. Worse, if one betrays the slightest sign of real interest in the place, serendipity takes over.

In the years that followed I visited every well-known, and some not-so-well-known places on the Colorado Plateau. I became irretrievably hooked. At every opportunity, I hiked, jeeped, flew over the canyonlands, and rafted down the Colorado. And, of course, my interest in photography was aroused by the compulsion to capture a moment, to explain a detail, and to share my experiences with others at home. Hearing of my infatuation and learning of my by then extensive 35mm collection, the Geological Museum in London proposed mounting an exhibit based on my Plateau photography. Then came one of those rare moments in life when a casual question becomes the fulcrum of one's destiny: Could I fill a large blank space over an auditorium in the Museum with a photographic mural of Grand Canyon?

I found no really informative books about panoramic photography. The nineteenth-century masters toyed with the idea of taking side-by-side multiple pictures to build up panoramas, but had problems with overlap, butting, image distortion, and fading at the edges of their prints. So they dropped this method, having concluded, one imagines, that the only solution must be one elongated photograph, for which they did not yet have the means. Hence the circuit camera, with its one-piece negative which reached its ultimate development in the 1920's. But there were snags about a camera that revolves about an axis while exposing a continuous strip of film. Landscape subjects were severely distorted unless the scene was either arranged in a semi-circle (like people in school photographs) or set at a very considerable distance from the camera. Then there was the fundamentally limiting factor of scanning everything horizontally. It seemed that I would first have to define "panorama" before I could begin. I decided that a view would only be panoramic if one had either to move one's

head up and down, or turn it from side to side, or both to take in the whole of it—if its total span was beyond human peripheral vision. None of the available panoramic cameras could produce a picture that fitted my definition. So I started where the early photographers stopped, by taking a series of discrete pictures that would be fitted together to make a whole. It proved to be a long, complex, and expensive road to travel.

The majority of people who visit Grand Canyon satisfy perhaps a lifetime's ambition when they do so. Many will never have the opportunity of seeing the Canyon at all. I kept my own first reaction to the Canyon very much in mind: How was the Canyon formed, how old is it, what do its spectacular features mean, and how do scientists work out the answers? I began by asking experts these basic questions. But as frequently as I obtained a reply more questions surfaced. I decided to keep to basic questions and to focus on the key subjects which I felt unlocked the meaning of the Grand Canyon region. And of course I tried to illuminate the major points with panoramas whenever possible.

As for my first panorama for the London Geological Museum, it was never used for its original purpose. My guess is that the person who asked the all-significant question didn't expect an answer.

Denver, Colorado
June 1980

The panoramic photography in this book is free of the distortion of distance normally caused by wide-angle lenses, or the curvature of nearby objects unavoidable with circuit cameras. As the diagram at right shows, the panoramas can be both horizontal, covering as many as 320 degrees, and vertical, taking in as many as 160 degrees above and below the camera viewpoint. No filters have been used to enhance colors, and the objective has been to reproduce landscapes just as they would be seen by the unaided eye. Each panorama is numbered and coordinated, and its location can be found on the foldout map on page 14. The angles of the panoramas are indicated by the following signs:

3. F-5.

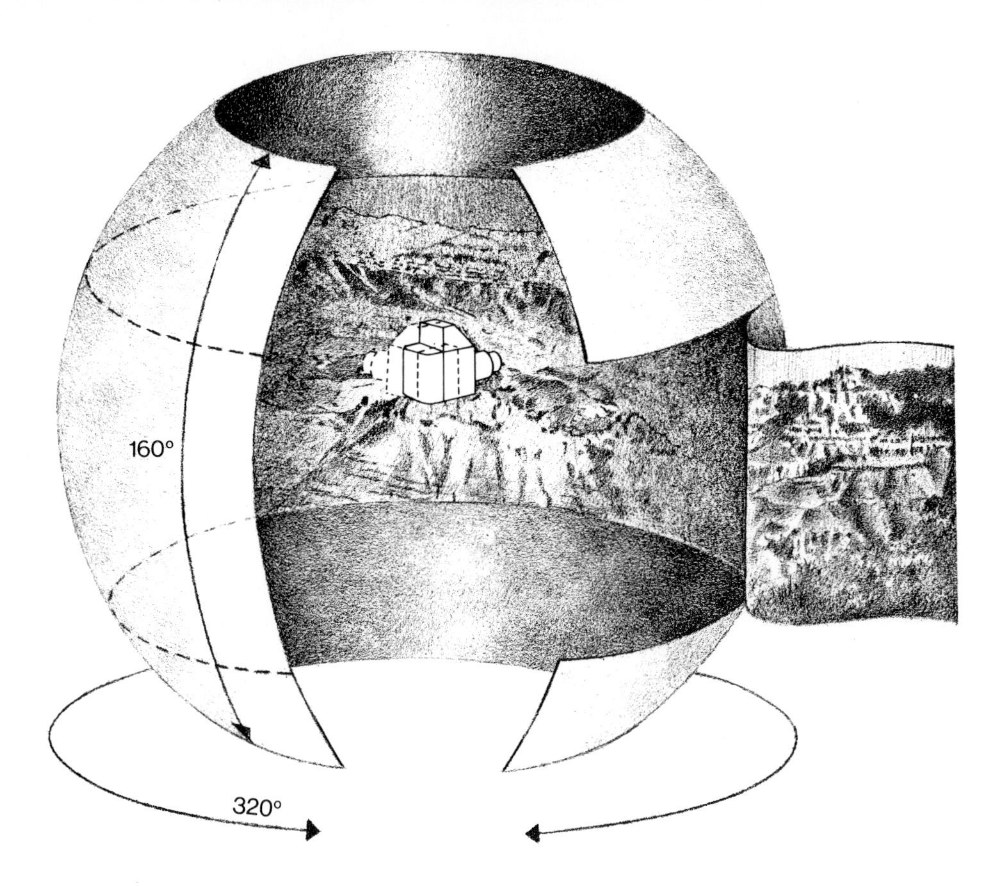

This cut from the panorama demonstrates the picture that could be taken from the same place by a 35mm camera with a 50mm lens.

I

THE NEW PANGENESIS

Upper Granite Gorge from Escalante Butte

The Colorado River to the left (west) has cut its way through rock formations a mile thick to reveal the geological story of Earth. Some of the oldest Precambrian sedimentary rock (ca. 800 to 1,200 million years old) in Grand Canyon is exposed at river level here, near Hance Rapid. The South Rim is at left, the North Rim is at right. The peak silhouetted against the North Rim near center-right is the Vishnu Temple (7,529-foot elevation).
2. F-5.

There is no greater contrast on Earth than that between the Grand Canyon of the Colorado and the Beardmore Glacier of Antarctica. The Grand Canyon is composed of mountain-sized buttes rising majestically from the floor of a vast canyon. It is predominantly red in color. The Beardmore Glacier divides the Transantarctic Range that straddles a continent almost buried in solid ice with immense plains of snow between its visible peaks. The whole scene is a blend of white and green and blue. In July of any year there can be a difference of two hundred degrees Fahrenheit between the depths of Grand Canyon and the slopes of the Beardmore Glacier. They are separated by great distance, about a third of the way around the globe. And yet these two places have a common ancestry, and their connection illuminates the story of Earth itself.

Early on a November day in the year 1912 the sun hovered low and large in a cloudless azure sky over the frozen desert of the Ross Ice Shelf. The bright sun saturated each crystal of snow so that the vast expanse scintillated in the still, crisp, subzero air. Spring had come and the slow Antarctic dawn had at last relieved the perpetual darkness of a winter-long polar night. A tiny fleck of color intruded into the otherwise pristine white wilderness. It was a green cone a foot or so high, visible for miles. The men who had traveled hard and far to look for it saw it, stopped in their tracks, hesitated, then trudged toward it with sinking hearts. They knew what they would find beneath thc cone: a small canvas tent that had become a tomb. The search party dug away the winter's accumulation of snow, packed hard against the tent walls, and found the entrance. Inside, Captain Robert Falcon Scott, the leader of the British

Polar Expedition of 1911, lay dead, half out of his sleeping bag, with one arm outstretched toward one of his two companions. His diary beside him told the whole tragic story of the disaster that had overtaken his party and of the selfless heroism to which he had been witness. The objective of Scott's party had been to be the first men to reach the South Pole. They succeeded in reaching it only to find that Scott's Norwegian rival Roald Amundsen had got there a month before them.

Edward Adrian Wilson, zoologist, artist, and chief scientific officer of the expedition, and Henry R. Bowers, two of the four men who had departed with Scott from base camp a year before, also lay as if peacefully asleep, Wilson with his arms crossed on his chest and Bowers laced up in a sleeping bag. All three had been severely frostbitten before they had mercifully succumbed to the deep slumber that precedes death from exposure. Scott's diary recorded that the two men who were missing from the original party that had set out from base camp a year before, Laurence E. G. Oates and Edward R. Evans, had died during the party's trek of eight hundred miles back from the Pole. One blizzard-torn day during this homeward haul, Oates, who had severely frostbitten feet, deliberately stumbled from the communal tent to his certain death in the biting wind. This ultimate sacrifice was intended to relieve his companions of the burden that Oates felt himself to be. Deep and bitter disappointment that they had not been first at the Pole, injury, exhaustion, and finally starvation contributed to their end. And now, here in a small tent, three of them lay as they had died eight months before, preserved by the freezing Antarctic night.

Buried under the snow a short distance away from the tent, the searchers found a sledge. Secured on it was a pile of rocks. Wilson had found them while doing some "geologizing" at the head of the Beardmore Glacier above the Ice Shelf three weeks before they pitched their final camp. Scott's diary for this particular day, February 8, 1912, read: "We found ourselves under perpendicular cliffs of Beacon sandstone, weathering rapidly and carrying veritable coal seams. From the last, Wilson, with his sharp eyes, has picked several plant impressions, the last a piece of coal with beautifully traced leaves in layers, also some excellently preserved impressions of thick stems, showing cellular structure." On the following day Scott added, "Wilson got great find of vegetable impression in piece of limestone."

Wilson had recognized the fossil rocks to be scientifically important. He was certainly able to identify one particular fossil embedded in them. It was of the genus *Glossoptera,* an extinct tropical fern. Its presence meant that the Antarctic continent, long barren of vegetation, had at one time been warm and certainly similar to regions of India and South Africa where fossilized *Glossoptera* had also been found. Many years before Wilson's find, the great Austrian geologist Eduard Suess had suggested a hypothetical grouping of the southern continents, basing his idea on the superabundance of the flora *Glossopteris* in the hemisphere. Suess coined the name "Gondwanaland" to identify the southern group and "Laurasia" for the lands of the northern hemisphere which had entirely different flora.

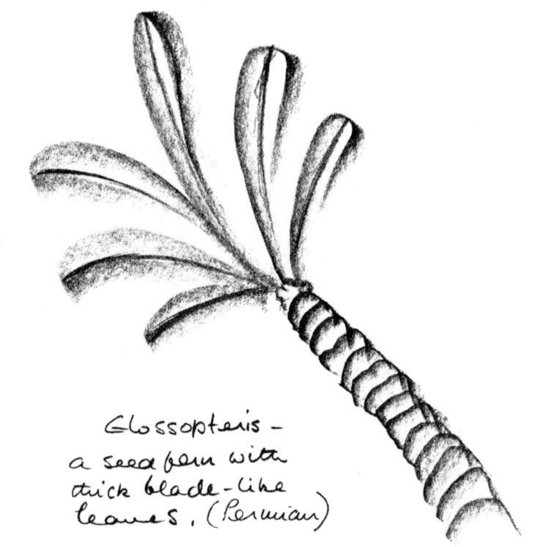

Marginal pencil drawings are by the author.

Wilson was more right about the profound significance of his find than he could possibly have guessed. Had their meaning been correctly interpreted in geological rather than botanical terms, the discovery would have been as scientifically important in its day as the return to Earth of rocks from the Moon has been in our own. The Moon rocks helped to establish the age of the solar system; Wilson's rocks could have played a key role in establishing the prior existence of a single ancestral supercontinent that split up to form the present seven continents.

By quite an extraordinary coincidence, at the very time that Wilson's specimens were lying on a sledge buried in Antarctic snow, an Austrian geologist named Alfred Wegener was giving lectures in Germany in which he hypothesized an original supercontinent, which he called "Pangea." His principal premise was the apparent jigsaw fit of the seven continents. In support of his theory he argued that the geology of corresponding coastlines was similar and that the distribution of ancient climates, as evidenced by fossils in rocks in now widely separated land masses, were incompatible with the present geographical positions of the continents. Wegener further argued that his concept was consistent with what was then known of the distribution of fossilized vegetation and animal remains. Wegener's theory was as revolutionary to geologists as the theory of natural selection had been to the biologists. Darwin sometimes referred to his theory as a "pangenesis." Perhaps Wegener's choice of the word "Pangea" (sometimes spelled Pangaea) was influenced by the Darwinian expression.

Others before Wegener had noticed the apparent intercontinental fit. For instance, Francis Bacon in his "Instauratio Magna" (a plan to reorganize the sciences) of 1620 made comments which are interpreted to be about the mirror image similarity between the coastlines of South America and Africa. Some went further and tried to fit the North and South American coastlines to corresponding parts of Europe and Africa. But until Wegener, none had had the temerity to suggest that the continents of today had split from a supercontinent, that they had separated by drifting over the face of the globe, and that they are still drifting.

In the years that followed his original pronouncements (published in book form in 1915) Wegener made a number of cardinal errors in the presentation of his ideas. He was also the strongest and most verbose advocate of his own theories, provoking a raging controversy that divided scientists into "drifters" or "antidrifters," and postponing a consensus for half a century.

Although the reality of continental drift had been reasonably well established in Europe and some other places by 1940, in fact as late as 1950 strong hostile opinions were still being voiced in the scientific world by those who preferred the "Isthmus Link Hypothesis"—continents joined by necks of land. But one comment made around that time typifies "antidrift" feelings. It was to the effect that Wegener's theories were "only a drunken sialic upper crust hopelessly floundering on the sober sima," implying that it was Wegener's mind that was wandering, not the continents.

Two further discoveries in the 1960's gained broad acceptance for the theory of continental drift. The first was that magnetized rocks can establish the position with respect to the poles of the continents on which they were formed ages ago. The second was that seabeds are spreading from central ridges on the ocean floors, and particularly that no seabed on earth appears to be as old as the continents themselves.

The Earth is polarized like a magnet, and when rocks cool from a very hot condition, any iron compounds that they contain naturally align themselves magnetically with the Earth's magnetic pole. The rocks' polarity remains permanently fixed, so that if rocks are moved, then their polarity will no longer be aligned with that of the Earth. Geologists found that they could identify misaligned rocks of the same age and

Escalante, early morning (gatefold)

The "engine room" of Grand Canyon. Taken from Escalante Butte, which lies some 2,000 feet below Lipan Point on the South Rim of Grand Canyon, this panorama takes in about 240 horizontal and 100 vertical degrees. So many geological events have occurred or are portrayed in this region of the Canyon that it warrants this description. It is authoritatively believed that between five and ten million years ago the Ancestral Colorado River flowed due south. During a period of uplift a second river cut west to east into the Kaibab Plateau—eventually "capturing" the southward main stream to form Grand Canyon as we know it today. But a billion years or more before these events the first series of sedimentary rocks in the Canyon were being formed, accumulating to a thickness far in excess of the present depth of the Canyon. Because of faulting, these rocks were tilted to form a mountain range and then eroded to a flat plain near sea level. These remnants of the oldest sedimentary formations in Grand Canyon can be seen stretching across the bottom two-thirds of the panorama from the left. They are characterized by the angle at which they lie and the horizontal surface to which they were eroded. There is a time gap of hundreds of millions of years between the top surfaces of the angled series and the undersurface of the horizontal rock formations which rest upon them. This time gap is an unconformity—the Great Unconformity of Grand Canyon. The first of these horizontal formations was laid down in a Cambrian sea about 550 million years ago. The rimrock which forms the horizon of the panorama was also formed under the sea, but over 300 million years later. Though there is little sign of it in these horizontal, undisturbed rocks, the supercontinent of Pangea was forming at the time they were laid down. The angular formations were deposited in seas covering an unknown configuration of continents which might also have formed a supercontinent a billion years ago. A reconstruction of the formation and subsequent separation of Pangea into the continents of today appears on page 23. 3. F-5.

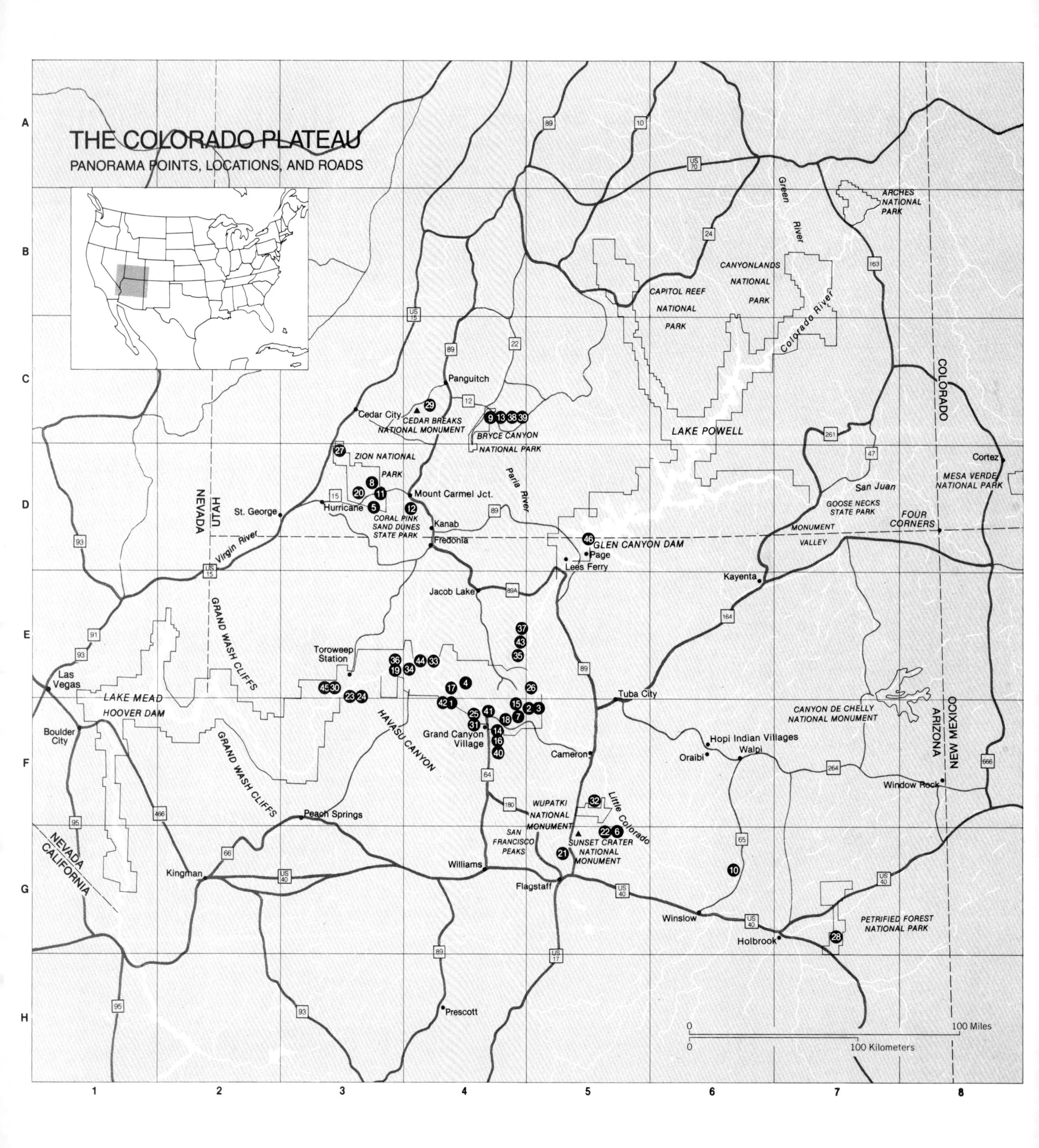
THE COLORADO PLATEAU
PANORAMA POINTS, LOCATIONS, AND ROADS
ARCHES NATIONAL PARK
Green River
CANYONLANDS NATIONAL PARK
CAPITOL REEF NATIONAL PARK
Colorado River
Panguitch
Cedar City
CEDAR BREAKS NATIONAL MONUMENT
BRYCE CANYON NATIONAL PARK
LAKE POWELL
COLORADO
Cortez
MESA VERDE NATIONAL PARK
ZION NATIONAL PARK
Paria River
San Juan
GOOSE NECKS STATE PARK
FOUR CORNERS
MONUMENT VALLEY
NEVADA
UTAH
St. George
Hurricane
Mount Carmel Jct.
CORAL PINK SAND DUNES STATE PARK
Kanab
Fredonia
Virgin River
GLEN CANYON DAM
Page
Lees Ferry
Kayenta
Jacob Lake
GRAND WASH CLIFFS
Toroweep Station
Las Vegas
LAKE MEAD
HOOVER DAM
Boulder City
HAVASU CANYON
Grand Canyon Village
Tuba City
CANYON DE CHELLY NATIONAL MONUMENT
ARIZONA
NEW MEXICO
Hopi Indian Villages
Oraibi
Walpi
Cameron
Window Rock
WUPATKI NATIONAL MONUMENT
Little Colorado
SAN FRANCISCO PEAKS
SUNSET CRATER NATIONAL MONUMENT
Peach Springs
NEVADA
CALIFORNIA
Kingman
Williams
Flagstaff
Winslow
Holbrook
PETRIFIED FOREST NATIONAL PARK
Prescott
0
100 Miles
0
100 Kilometers

THE COLORADO PLATEAU
PRINCIPAL PHYSICAL FEATURES
N
PINE VALLEY
Meadow Valley Wash
MORMON RANGE
Virgin River
SHIVWITS PLATEAU
UINKARET PLATEAU
HURRICANE
TOROWEAP VALLEY
Whitmore Wash
KANAB PLATEAU
Kanab Creek
KANAB CANYON
MARKAGUNT PLATEAU
PAUNSAUGUNT PLATEAU
AQUARIUS PLATEAU
PINK CLIFFS
GRAY CLIFFS
CLIFFS
PARIA PLATEAU
Paria River
VERMILION CLIFFS
KAIBAB PLATEAU
MARBLE CANYON
ECHO CLIFFS
KAIBITO PLATEAU
KAIPAROWITS PLATEAU
GLEN CANYON
LAKE POWELL
HENRY MTS.
Fremont River
Dirty Devil River
SAN RAFAEL SWELL
GREEN DESERT
Green River
Colorado River
BOOK CLIFFS
San Juan River
LAKE MEAD
GRAND WASH
GRAND CANYON
MIDDLE GRANITE GORGE
LOWER GRANITE GORGE
UPPER GRANITE GORGE
BRIGHT ANGEL CANYON
HUALAPAI CANYON
HAVASU CANYON
HUALAPAI PLATEAU
GRAND WASH CLIFFS
HUALAPAI VALLEY
BLACK MOUNTAINS
AUBREY CLIFFS
COCONINO PLATEAU
GRAND FALLS
SAN FRANCISCO PEAKS
SUNSET CRATER
Little Colorado River
PAINTED DESERT
MOGOLLON RIM

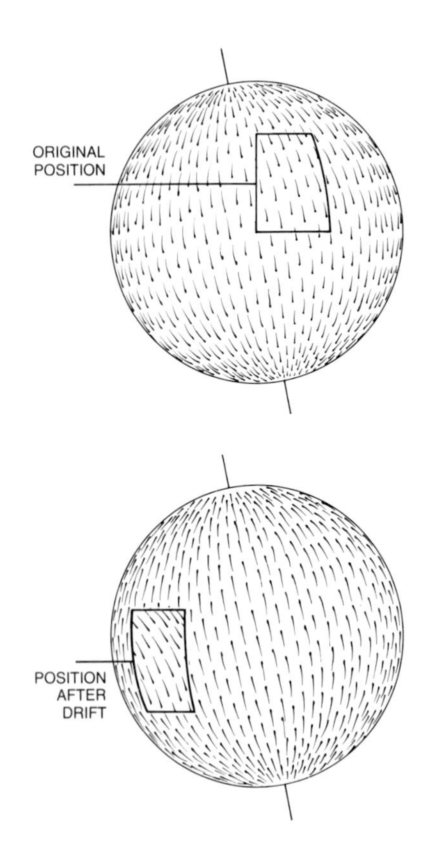

The Earth's magnetic field is polarized north-south, but the direction of the magnetic force has frequently reversed itself. Rocks containing iron have magnetic "fingerprints" which record the direction of the magnetic force prevailing at the time they were formed. When such "fingerprints" are plotted from many samples the original location of the continents on which they were formed can be calculated.

similar magnetic orientation and could reconstruct their original position in relation to the magnetic pole. From this they could deduce the different positions of the continents at the time the rocks took on their magnetic alignment.

Reaching this conclusion had been fraught with difficulties. That certain rocks become magnetized when formed had been discovered in classical times. In the early nineteenth century it was found that some rocks appear to be reversely magnetized, that is, they are polarized in the opposite direction to the present location of the magnetic pole. This phenomenon was not understood until the 1950's and its significance not established until 1963, by which time it had been discovered that rocks capable of being magnetized, from all regions of the Earth's surface, had been reversely magnetized on so many different occasions that in sum total they had been reversely magnetized during approximately half geological history, and normally magnetized during the other half. It was concluded from this that the magnetic pole of the planet Earth switches unpredictably from north to south and vice versa, and that the period between switches lasts many thousands of years, leaving a magnetic record in the rocks. These determinations contribute to the reconstruction of the course of continental movement.

Because about seventy percent of the Earth's surface is under water, much of it cannot be observed directly. Oceanographers had been mapping the seabeds with increasing intensity with equipment developed in World War II. They found great ridges on all the ocean floors, which seemed to be interconnected—but they were not sure of their significance. In the 1950's scientists from the Scripps Institution of Oceanography towed a magnetometer back and forth over a large area of the eastern Pacific which was ridged in this way. They discovered long, narrow parallel bands of the seabed, aligned and magnetized alternately, north and south. Other scientists then found the same phenomenon in other ocean basins. Two Cambridge University scientists, Fred Vine and Drummond Matthews, proposed that the parallel magnetic striping was the result of the magnetization of magma as it emerged from the mid-ocean ridges: when there was a pole reversal the magma would be magnetized in the opposite direction. In 1965 scientists at the Lamont Geological Observatory found a South Pacific magnetic profile in which a series of magnetically alternating bands stretching west from the mid-ocean ridge perfectly matched the series stretching east from the ridge. They were mirror images. This was the conclusive evidence for the seafloor-spreading hypothesis: all the ocean beds on earth are flowing outward from central ridge-rifts toward the continental masses. It was no wonder that the ocean beds appeared to be younger than the continents, for they are being formed on a day-to-day basis. Soon after it was established that the spreading ocean floors are being consumed or destroyed at some continental margins. To understand the possible driving forces of this process, the "tectonic" forces, we must briefly consider the current theory about the formation of the solar system and so of planet Earth.

It is considered most likely that the formation of the whole solar system began as a process of intense gravitational attraction of interstellar gas and dust revolving about a central pivot, in overall shape not unlike a hurricane. This nebulous disk evolved with the proto-sun at its center and with the nine planets, still in the form of bands of particles. The disk was hotter at its center than at its extremity and, as it rotated, individual spheres began to concentrate by force of gravity. Like skaters on an ice rink, the more compact the spheres became, the faster they rotated and, conversely, the faster they spun around their own axes, the more compact they became. A tremendous gravitational force developed, causing the collapse of the proto-sun into a star, and the compression of the forming planets into dynamic spheres with intensely hot liquid cores.

Hot particles in a semiliquid mass expand and rise, while cooler particles contract and sink. Such movement is called "convection." The cooling of the Earth's surface by radiant-heat loss and the generation of heat in the interior by a combination of pressure and radioactive decay caused convective movement within the sphere. Heat loss was greater than heat production, so Earth cooled, and, as it did so, the heaviest elements began to solidify. Iron and nickel gravitated to the center to form a core, while granite, a composite of comparatively light elements, rose like impurities on the surface of molten metal to the surface to form islands, and eventually the first continental masses.

This explanation serves to illustrate why continents of granite, a comparatively light form of rock which is rich in silicon and aluminum (the "sial"), float on a mass of dense basalt which is rich in iron and magnesium (the "sima"). These layers of the Earth's crust are known respectively as the "continental crust" and the "oceanic crust," and in combination with the upper mantle form a rigid envelope termed the "lithosphere." The immediately underlying part of the mantle is plastic and is called the "asthenosphere." The driving force of seafloor spreading which causes the continents to "drift" is considered by many geologists to be derived from the convective movement of the asthenosphere (see illustrations pages 22–23). The whole complex process is given perspective by imagining that if the Earth were to be reduced in size to a globe ten feet in diameter, the average thickness of the continental crust would be three-quarters of an inch, and the oceanic crust one-tenth of an inch.

The surface of the earth is made up of plates consisting of oceanic crust within which the continents "float" (see illustration page 23). Ridge-rifts are mid-ocean boundaries between plates from which fresh basaltic magma from the asthenosphere is continually being exuded. Imagine dense molten magma oozing from the full length of a central ridge-rift which then cools and solidifies. After a period of time, a further wave of molten material oozing from the ridge-rift will in effect divide the first accumulation into two parts. The bands of older rock on either side of the new emanation, while remaining parallel, will move away from each other, assisted by convective movement of the asthenosphere. As the spreading plates grow in size they tend to push their continental passengers in the direction of the spread. Because of this the continental plates tend to move either away from each other, as in the case of the African and North American continents with the Atlantic Ocean widening between the two, or toward each other like the subcontinents of India and China. The inevitable collision of such continental masses moving in opposition leads to the formation of a mountain range, in this case the Himalayas. In some instances, plates converge at an angle or slide past each other and areas of a continental mass are carved off the mainland. The Baja California peninsula and adjacent parts of the California coast are examples of a pending event of this kind.

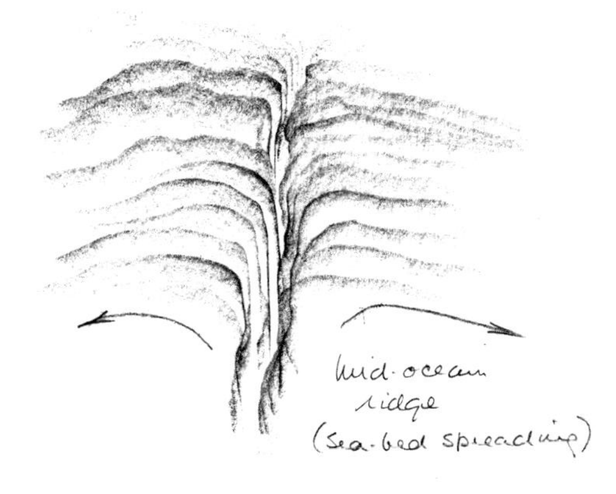

But if all the plates were simply to expand ad infinitum, the earth itself would have to expand to accommodate them. Some scientists still consider this a serious hypothesis, but what we do know is this: one interacting plate simply dives under

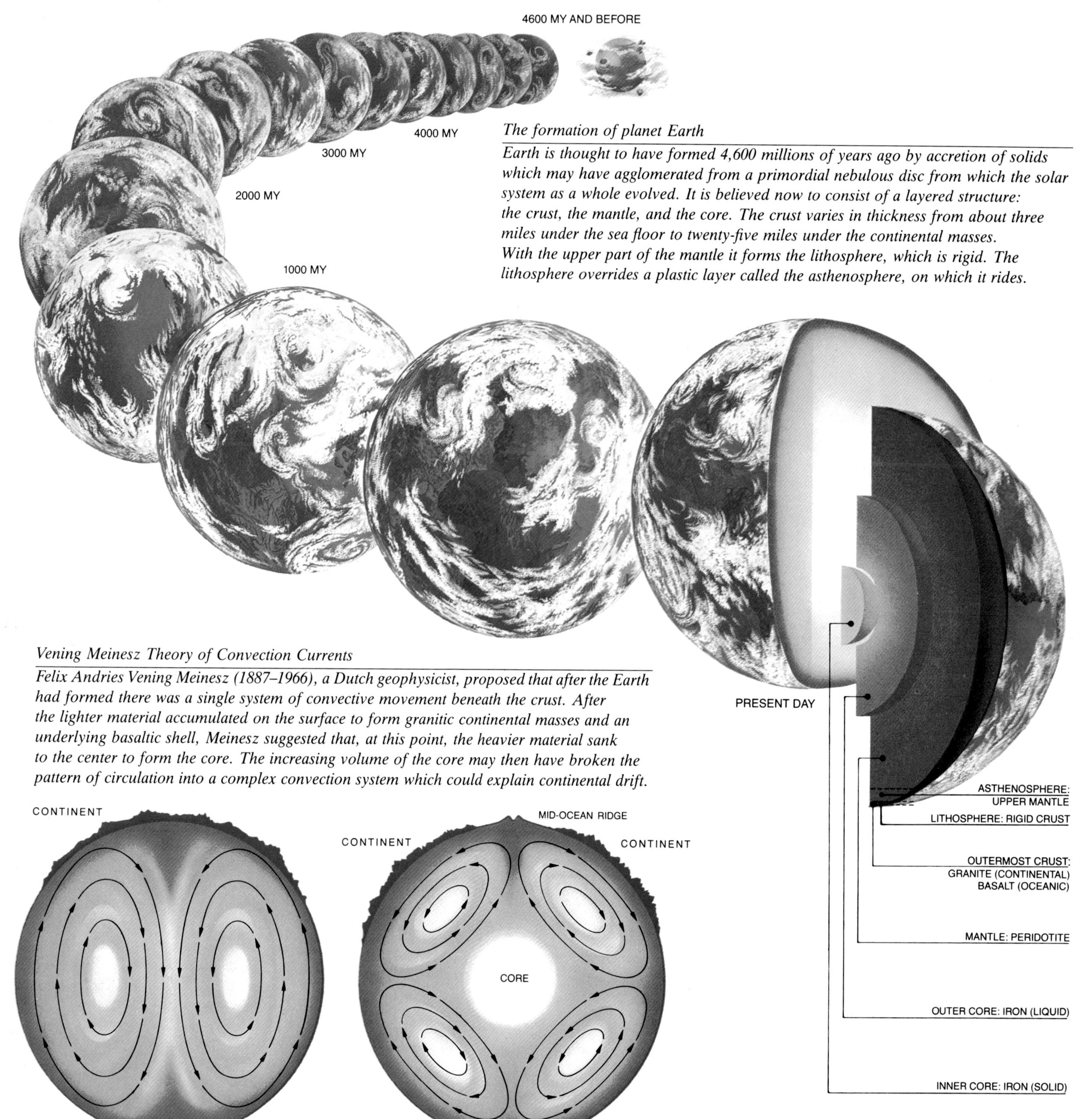

The formation of planet Earth

Earth is thought to have formed 4,600 millions of years ago by accretion of solids which may have agglomerated from a primordial nebulous disc from which the solar system as a whole evolved. It is believed now to consist of a layered structure: the crust, the mantle, and the core. The crust varies in thickness from about three miles under the sea floor to twenty-five miles under the continental masses. With the upper part of the mantle it forms the lithosphere, which is rigid. The lithosphere overrides a plastic layer called the asthenosphere, on which it rides.

Vening Meinesz Theory of Convection Currents

Felix Andries Vening Meinesz (1887–1966), a Dutch geophysicist, proposed that after the Earth had formed there was a single system of convective movement beneath the crust. After the lighter material accumulated on the surface to form granitic continental masses and an underlying basaltic shell, Meinesz suggested that, at this point, the heavier material sank to the center to form the core. The increasing volume of the core may then have broken the pattern of circulation into a complex convection system which could explain continental drift.

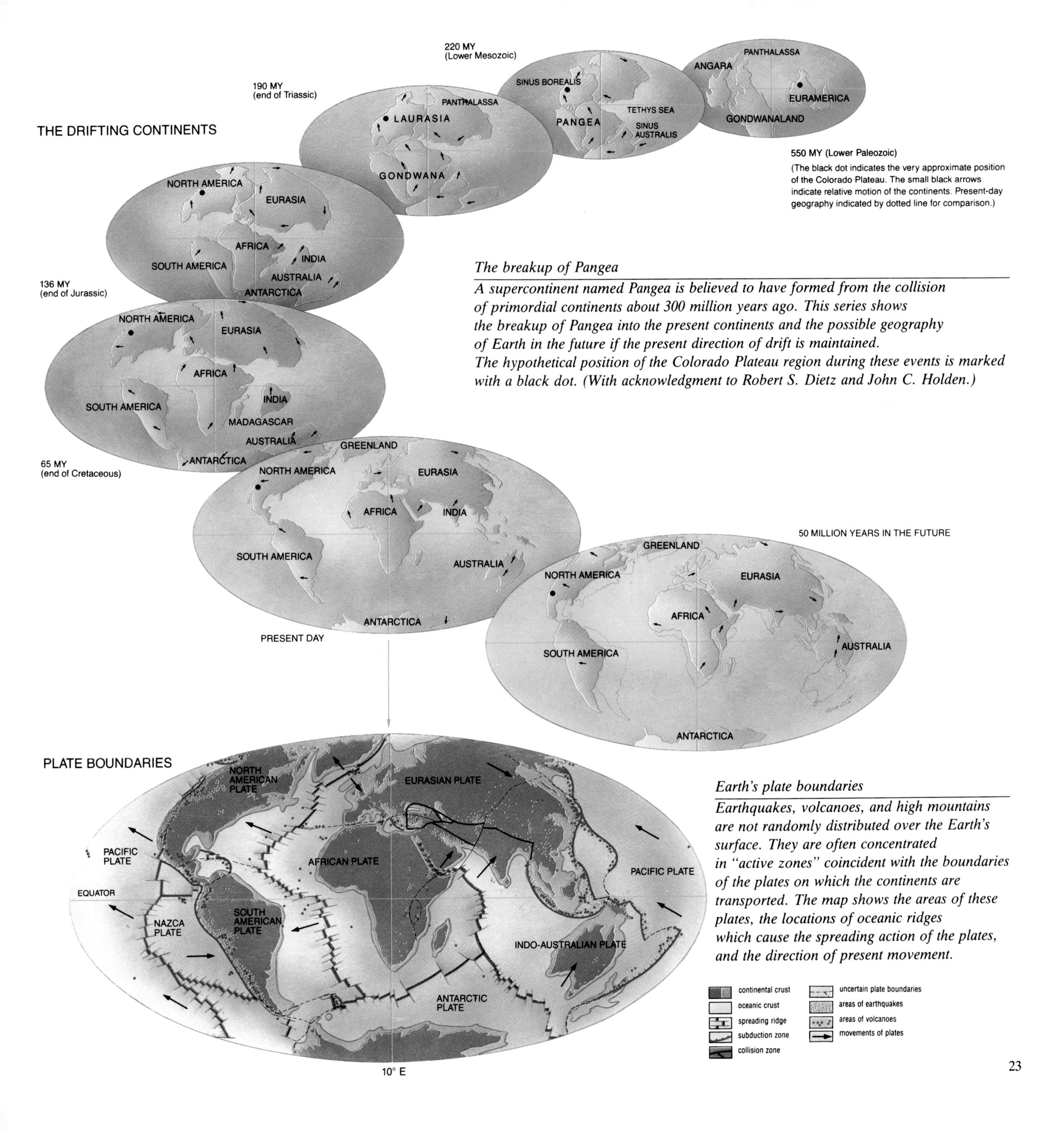

The breakup of Pangea

A supercontinent named Pangea is believed to have formed from the collision of primordial continents about 300 million years ago. This series shows the breakup of Pangea into the present continents and the possible geography of Earth in the future if the present direction of drift is maintained. The hypothetical position of the Colorado Plateau region during these events is marked with a black dot. (With acknowledgment to Robert S. Dietz and John C. Holden.)

Earth's plate boundaries

Earthquakes, volcanoes, and high mountains are not randomly distributed over the Earth's surface. They are often concentrated in "active zones" coincident with the boundaries of the plates on which the continents are transported. The map shows the areas of these plates, the locations of oceanic ridges which cause the spreading action of the plates, and the direction of present movement.

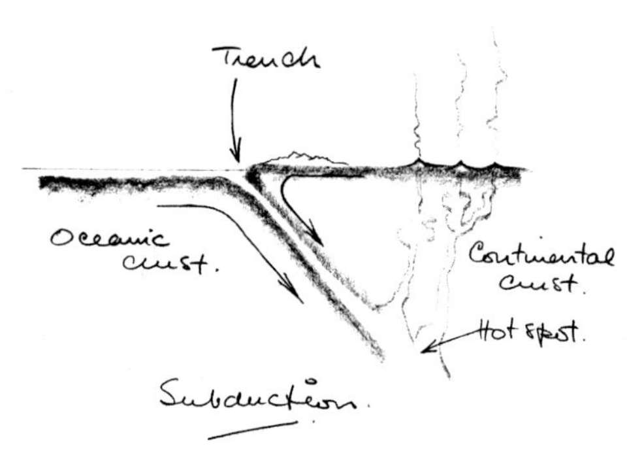

another, a phenomenon called "subduction." Where the Nazca Plate plunges under the South American Plate, subduction produces earthquakes, volcanoes, and mountain ranges for the whole length of the west coast from Ecuador to Chile. The subducted plate plunges downward at an angle, to be melted and reabsorbed into the mantle.

The picture of the Earth that has gradually emerged is one of a dynamic upper crust which consists of comparatively light and rootless continents which float helplessly on a high-density substratum in constant motion. One might suggest, in fact, that that rather caustic description of Wegener's theory, "a drunken sialic upper crust hopelessly floundering on a sober sima," was nearer the mark than its author intended.

From the beginning of all the theorizing, through all the controversy, and even after the physical evidence had accumulated massive support for the theory of continental drift, it was still held by many scientists that the role of Antarctica in this context was the vital one. It was there, at the bottom of the world, from the rocks piercing the snow and ice mantle of that isolated and frigid continent, that the final conclusive evidence would have to come. The antidrifters reasoned that Wilson's fossil flora *Glossopteris* could have been born of seeds carried by the wind, by sea currents in times of a warmer Antarctic climate, or by land by means of a connecting isthmus. If the massive Antarctic continent had indeed drifted into its present position from its supposed original location in southwestern supercontinental Pangea, then land animals which pre-dated continental drift would be found as fossils in both continents. In the absence of connecting isthmuses thousands of miles long, there is no way that the same form of animal land life can appear locally and spontaneously on a number of continents.

Until the late 1960's no evidence of such animal life form had been found in Antarctica. And then, one day late in 1967, a young geologist named Peter J. Barrett of the Institute of Polar Studies, Ohio State University, found a fragment of what he thought was fossilized bone while prospecting a region near the Beardmore Glacier. The specimen was sent for examination to the American Museum of Natural History in New York, where one of the world's leading paleontologists, Edwin H. Colbert, identified it as a small section of the jawbone of a very early amphibian. The crucial and exciting point was that the particular animal that Dr. Colbert believed to be the original owner of the jawbone was a freshwater amphibian and could not have swum in the sea.

This startling development led to a change in emphasis for geological research in the south polar region. Consequently, in November 1969, Colbert and a team of fossil hunters were airlifted to a research base set up west of the Beardmore Glacier near Coalsack Bluff, a mountain piercing the ice. The intention was to have U.S. Naval helicopters drop the party on Graphite Peak, where the original fragment of bone had been found. Weather made helicopter flights impossible, so the leader of the party, David Elliot, together with James M. Schopf, a specialist in fossil flora, decided to investigate the more accessible Coalsack Bluff, since it appeared to be similar in geologic structure to Graphite Peak.

One can imagine the excitement when, almost immediately after arrival on the bluff, pieces of what appeared to be fossilized bone were found. There was a dash back to base camp on the motor sledge to obtain Colbert's opinion and the eventual return of the whole paleontological party to the locality of the find when he had pronounced affirmation: Yes, they most certainly were bones. In the weeks that followed four hundred and fifty specimens were collected. It was a chance find in a million. And events had turned full circle. Coalsack Bluff was near the very place on the Beardmore Glacier where Wilson had picked up his fossil rocks fifty-seven years before.

Within days of the first fossil discovery, Colbert knew that the final piece of evidence for the theory of the prior existence of Gondwanaland, and therefore of continental drift, had been found—the incontrovertible piece. Many of the fragments of fossil bone were from a species of dinosaur called *Lystrosaurus* and one, an upper jaw fragment, was complete with tusk. In life *Lystrosaurus* was small, seldom more than two feet long. It was a freshwater amphibian and a herbivore. It lived about two hundred and fifty million years ago and was prolific in regions of South Africa and India. On examination back in the United States, the Antarctic fossil specimens proved to be a perfect replica of the African and Indian forms.

The Coalsack Bluff *Lystrosaurus* could not have swum the oceans to Antarctica. It could not have crossed imaginary isthmuses to get there. Neither could it have flown, nor could it have evolved spontaneously. The only plausible reason for its presence was that at one time South Africa, India, and Antarctica had been joined. *Lystrosaurus* was the missing link in the long chain of hard evidence that geologists needed to establish the pre-existence of a primeval supercontinent. This, an historical fossil find, was one of the final pieces of the jigsaw that completed the picture. The great supercontinent that Wegener named "Pangea" really had existed.

Perhaps the best place on Earth to contemplate the extraordinary sequence of events which led to this conclusion is on the edge of the South Rim of Grand Canyon, Arizona. There, lying below and spreading out on either side to a degree beyond human peripheral vision, is a prospect difficult to imagine. Before you lies a section cut through part of Pangea; a glimpse of the primordial supercontinent itself. And much more than that, for the walls of Grand Canyon preserve the record of a great many of the momentous geological events that preceded the formation of Pangea.

Grand Canyon, neighboring Painted Desert, and Zion and Bryce canyons demonstrate different stages in the formation and eventual disintegration of the last supercontinent—"last" because there most probably was a supercontinent before Pangea and there will probably be others in the future. The rocks that can be seen on the Colorado Plateau today were formed at distances up to a thousand miles or more from their present position.

In 1969, while Dr. Colbert and his associates were chipping away at Coalsack Bluff, I was taking my first look at Grand Canyon from the popular South Rim along with several million other people who did the same thing that year. Little did any of us realize it, but those paleontologists working in the bitter gusting winds of Coalsack Bluff were not only making a most significant fossil find, but in a sense they were determining how we visitors should look at that bewildering Canyon world.

When I first looked into the chasm, I wondered unbelieving at the incredible depths, marveled at the violent thunderstorm that was dramatizing the scene above and below the North Rim opposite, made a few wild and very inaccurate guesses about how the Canyon was formed, and within twenty-four hours left, thinking I would never see that world wonder again. Typical. Now, ten years and many visits, hikes, helicopter

landings, and river trips later, I see the Canyon in an entirely different way, much influenced by scientists like Dr. Colbert.

Today, when I look into the depths of Grand Canyon at rock formations layered one upon the other, I imagine the primeval environments that existed when each in turn was top of the pile. And when I look at the rocks of Granite Gorge in the very bowels of the canyon, I visualize the forming seas two thousand million years before their time, the continuous rain for centuries on end, cloud cover as black as night with perpetual lightning, and belching volcanoes to provide the only illumination. A minute upward movement of the eye spans an inconceivable length of time to a rock formation, once the bed of a tropical sea, now frozen in time with its marine life fossilized, still, waiting for a paleophile to find it, and wonder at it. All these events are really the functions of unimaginable, endless, unremitting, and uncompromising time. How, I wondered, do geologists organize their thinking about time?

A Scottish geologist named James Hutton first formally proposed the principle

From Point Sublime, Grand Canyon, North Rim

With the exception of the near vertical cliffs of schist and granite in Granite Gorge, at the top center of this panorama, the rocks in view were formed from seabed sediments, and in estuarine and desert environments during the Paleozoic Era (570 to 225 million years ago). This view from Point Sublime was immortalized by W.H. Holmes, who drew a panorama of it for reproduction in Clarence E. Dutton's report for the U.S. Geological Survey of 1880–1881 (see pages 168–169). 4. F-4.

of classification of rocks into strata in the late seventeenth century. Subsequently, an English civil engineer, William "Strata" Smith, who collected fossil shells as a hobby, noticed that some layers of rock contained fossils which differed from those in neighboring series. This discovery led to the development of a catalog of index-fossils with which scientists are able to allocate rock formations in various parts of the world to particular divisions of time on the basis that the oldest rocks, and therefore the oldest fossils, are at the bottom of a series. As fossil evidence accumulated, rock series were allocated to particular "eras," and eras were further classified into "periods" of time. There are four major geological eras. They are called the Cenozoic, age of recent life; the Mesozoic, age of middle life; the Paleozoic, age of ancient life; and the Precambrian, the age from the formation of Earth to the appearance of abundant life. The Precambrian Era represents about eighty-five percent of the total age of the planet and is normally divided into the Archean, for the oldest known rocks and unicellular organisms, and the Proterozoic, for the oldest known multicellular organisms.

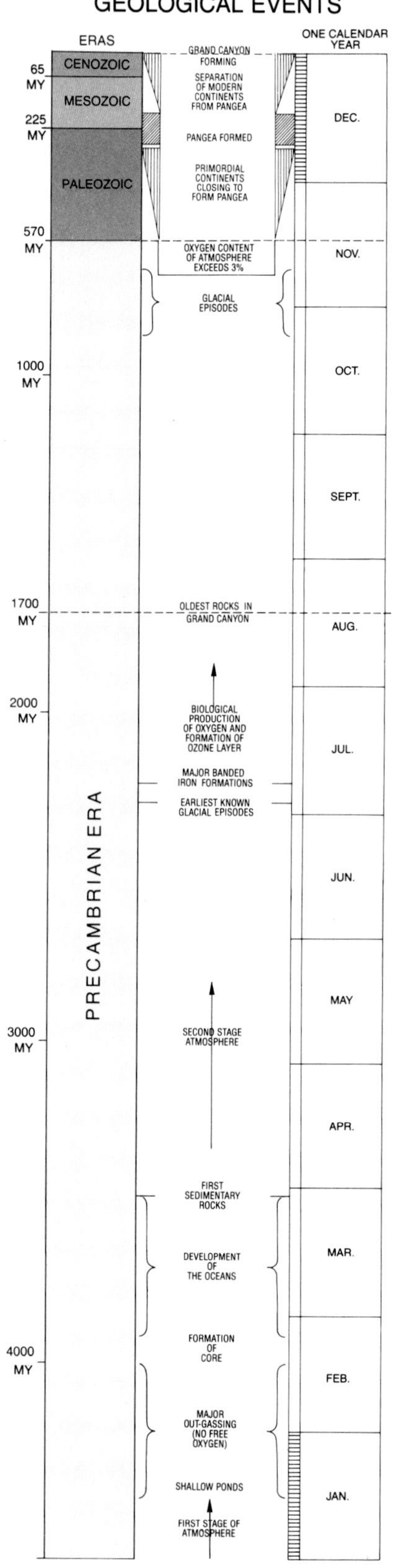

The periods allocated to each era differ in both number and length of time represented. They start with marine invertebrates in the early Paleozoic, called the Cambrian Period because the first evidence of complex forms of life was found in Wales, known to the Romans as "Cambria." The Ordovician and Silurian Periods of the Paleozoic Era are named after Celtic tribes, the Silures and the Ordovices, indigenous to Wales during the Roman Conquest. These two periods witnessed the development of the first fish and the first air-breathing invertebrates, the millipedes, and scorpions. The Devonian Period which followed is known as the Age of Fish because of the predominance of fish among living creatures at that time. The derivation of the term "Devonian" is also British, the rocks of the county of Devonshire in southern England being the first to yield clues to this period of the distant past.

The transition of geology from its early stages as an intellectual pursuit for talented British amateurs into an international science is characterized by the derivation of names for the remaining three periods of the Paleozoic Era in the United States, which differ from those used in Europe. These start with the Mississippian, after the Mississippi River valley, the period during which the first winged insects appeared; Pennsylvanian, after the state of Pennsylvania, when prolific carboniferous forests existed; and Permian, after the town of Perm, USSR, where evidence of the first quadrupeds was found.

The Mesozoic Era is divided into three periods: Triassic, of German origin, denoting the three characteristic types of rock typically associated with the period; Jurassic, after the Jura Mountains of Switzerland; and Cretaceous, after the Latin word for chalk and the chalky white seacliffs of southern England. It was during these periods that the dinosaurs developed, dominated, and became extinct.

The Cenozoic, the present era which began in the relatively recent past (about sixty-five million years ago), is divided into two periods, Tertiary and Quaternary (terms left over from an earlier system). Because there is such a vast accumulation of geological facts relating to these last two periods, they have each been subdivided into seven "epochs," covering the emergence of mammals as the dominant form of life to the appearance of man and to the present day.

Thus eras are divided into periods and periods are subdivided into epochs. Their nomenclature is used throughout the world by earth scientists. The names given to individual rock formations, however, are applicable only in a particular locality, because local rock formations may differ substantially in character from rocks of the same period, whether a few hundred or many thousands of miles away. One locality

The time scale of Earth's evolution

The diagram on page 29 illustrates the time scale of life's development,the one at left the time scale of some of the physical events which coincided, as far as is known.
The scale condenses the passing of the 4,600 million years of Earth's supposed existence into one calendar year—the equivalent of two lengthy human lifetimes per second.
Thus, on this scale, the known history of the genus ‹‹Homo›› (10 million years) occupies little more than half of one day—December 31—and the whole of the history of multicellular organic life on Earth is covered by the months August to December. First life is believed to have evolved as early as the middle of February. The oldest rocks in the Grand Canyon (Vishnu Schists) were formed during late July and early August, and the earliest sedimentary rocks were formed in the primeval seas during February.

may have been a river delta and another the region of a shallow sea, thus creating conditions for the formation of different types of rocks in the same period of time. For this reason it is customary to name a formation after a locality, usually the place where the formation was first identified.

No matter how confusing this system appears to be to the layman, or how onerous to the student of geology who has to learn, mark, and inwardly digest literally hundreds of strange-sounding names of formations around the world, logic does apply. It is only necessary for a geologist to identify the period or epoch from which a particular sample of rock has been obtained, for him or her to be understood when discussing that specimen of rock with any other geologist.

The oldest rocks in Grand Canyon are now thought to be approximately one billion seven hundred millions of years old. The Earth was formed about four billion six hundred millions of years ago. How does one begin to comprehend time on this enormous scale? The allocation of time to eras and periods, each of multimillions of years in duration, is all very well, but many people find it difficult, if not impossible, to visualize even one million years, which is no time at all in geological terms. To get a sense of how big the number "one million" is, one wonders how long would it take to count? A pocket calculator provides the answer; it would take about a month, intoning constantly for eight hours each day, seven days a week. What about counting the number one hundred million? Obviously, about one hundred months—a little more than eight years. Assuming a long life, anyone starting to count during infancy could count up to a billion if he counted for the rest of his life!

Instead of wasting a metaphorical lifetime counting improbable numbers, suppose that the whole extent of one human lifetime could be compressed into just half of one second of normal time. Just imagine all the experiences of a seventy-three year old from babyhood to old age condensed into a fraction of a second. On this basis, the Egyptian boy-king Tutankhamen was buried in his golden sarcophagus less than a minute ago; the first humanoids put in an appearance just nineteen hours ago, about the same time that Grand Canyon began to form. The primeval continent of Pangea began to break up into the seven continents we know today about sixteen days ago. And one calendar year would represent the passing of four billion six hundred million years, the approximate age of Earth.

The object of this analogy is to persuade the reader to abandon normal time perspective when reading this book. Sense of time is directly linked to the daily act of living. To us humans, rocks are solid, immovable, almost indestructible, although in fact, geologically they are fluid, fragile, and constantly regenerating themselves. One of the ways in which we can imagine rocks behaving in the way they actually do behave is to visualize the whole geological process taking place at a rate equivalent to several human lifetimes for each breath we take.

In the context of geology, geophysics, and associated earth sciences, our species, *Homo sapiens*, is a microorganism in a plastic world. In the perspective of infinite time, we humans are an obscurity in a dynamic environment, a mere aberration, unfelt, ineffective, and almost undetectable. Just as we need time-lapse film sequences to understand how a rose unfolds its petals, so we must adjust our imagination to accept the extent of our planet's potency, so vividly illustrated in the canyons of the Colorado Plateau.

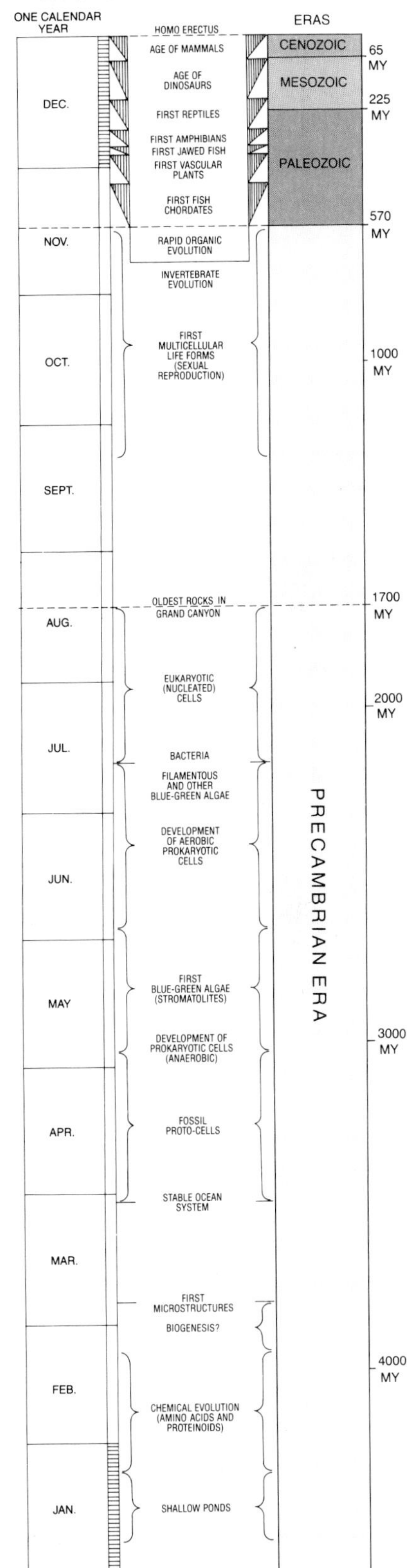

II
CORRIDORS OF TIME

The Rockville Vista

The portals of Zion Canyon caught in the October evening sun. Mt. Kinesava dominates the center left, and the East Temple is at center right. The Moenkopi Formation at lower left covers the Kaibab Limestone, which forms the rims of Grand Canyon. 5. D-3.

A towering neck of sculpted basalt, the eroded remnant of a volcanic pyramid of a bygone age, stands in solitude near the northwestern limit of the state of New Mexico. For fifty miles around it the country is arid, barren, undulating desert. The only other remarkable object in its immediate vicinity is a large, flat, pink-granite plinth. The plinth is capped with an appropriately engraved bronze plaque boldly proclaiming that this is the one place in the United States of America where four states meet at a single point. By standing on the center of the plinth facing north, with the right foot inside New Mexico and the left inside Arizona, and by placing the right hand in Colorado and the left in Utah, it is possible to achieve the distinction of being in all four states at the same time. During the summer months hundreds of visitors perform this ritual while Navajo Indians around the shrine sell bead necklaces to the worshipers.

But the monument has another and more important significance. It is the symbol that identifies a large segment of the southern province of the Colorado Plateau referred to as "Four Corners Country." The Arizona-Utah state line runs due west from here and, after crossing one hundred and fifty miles of desert, it divides the regions of Grand Canyon and the Painted Desert in northern Arizona from those of Zion and Bryce canyons in southern Utah. From whichever direction the three canyons are approached, the general inclination is uphill; from Phoenix, Arizona; from Las Vegas, Nevada; from Grand Junction, Colorado; or from Four Corners. In fact, the uphill trend from east to west begins at the Mississippi River, with what geologists call a "gentle warp," and ends rather abruptly at the western extremities of the uplift where the Colorado Plateau is confined by the Basin and Range Province (see illustration pages 32–33). If it were not for the upwarp there would not be a Grand Canyon, or a Zion, or a Bryce Canyon, and the whole area of the Painted Desert might very well be at the bottom of a vast lake fed by the Colorado River.

Between ten and twelve million years ago, when humanoid creatures were beginning to learn how to fashion tools and to stand naturally upright, the North American continent was one hundred or so miles east of its present position, and nearing the end of its long westward drift to its present geographical position. Then the region now called the southern province of the Colorado Plateau began to uplift at a rate averaging one hundredth of an inch a year; it continued to do this for about five million years. The resultant increase in elevation of more than four thousand feet had a devastating effect.

The cause of the Colorado Plateau uplift has always been the subject of intense debate among geologists. They agree that crustal instability caused by "tectonic" forces, the forces that deform the Earth's crust, was responsible; the problem is to determine which tectonic force did what. Some think that the movement of the North American continental plate from east to west, driven by the seabed spreading of the mid-Atlantic ridge, could not have made a contribution; the Colorado Plateau is just too far away. On the other hand, the seabed of the Pacific Ocean spreads just like the Atlantic seabed, but the difference is that where the Pacific plate meets the North American plate, it descends below, creating a subduction zone in which the rock of the Pacific plate is reabsorbed into the Earth's asthenosphere. As passengers on the Pacific plate came a jostling pack of small "microcontinents." The mountain ranges which stretch parallel to the Pacific coast of North America, from Alaska in the north to the Baja peninsula in the south, are the result of multiple collisions of these microcontinents as they piled up against the old core of the continent.

The presence of a subduction zone under the western edge of the North

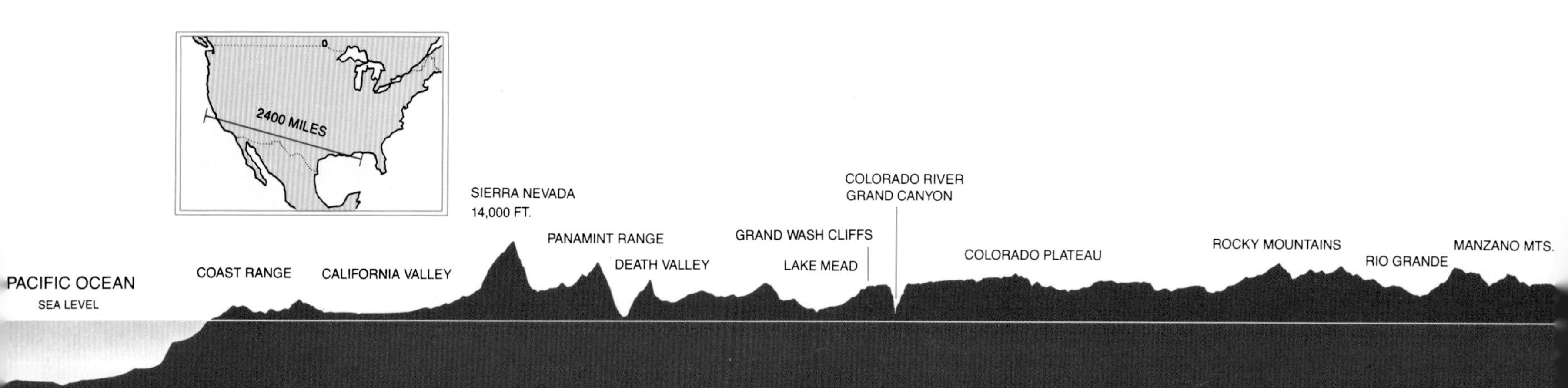

American plate, which provides an apparently obvious answer to the question of what caused the Colorado Plateau uplift, is not considered a reliable solution for several reasons. First, at some undetermined time, the Pacific plate appears to have changed its direction of movement northward. If there ever was a head-on collision of plates, it certainly is no longer the case. Today it is more of a glancing blow which is demonstrated by the development of the San Andreas fault line and the consequent tendency of a large section of the North American plate to be sheared off into the Pacific Ocean. Second, no one knows at present at what angle or to what depth the Pacific plate plunged under the North American plate and therefore the extent to which subduction influenced the uplift of the Colorado Plateau. While it is reasonable to suppose that subduction did make a contribution, it may have only been an indirect one. But it is certain that another process called "isostasy" did play an important part.

The theory of isostasy is based on the fact that mountains and lakes and other heavy natural masses on the Earth's surface exert a downward force that is proportionate to their density and size, and thus cause a temporary sinking of the Earth's crust which is relieved only when the object is removed. Perhaps the best way to visualize the mechanical interaction between the continents and the Earth's crust is to substitute a water bed for the crust and a large flat chunk of wood for the continent. The wood will sink down into the water bed until it is supported by the water confined in the plastic skin of the bed. The bed's surface tension will increase and its overall surface will rise to compensate for the wood. If the wood is removed, surface tension is reduced and the water bed will resume its original level.

The uplift of the Colorado Plateau was influenced by the factors illustrated by this analogy. For example, the formation by faulting of the Basin and Range Province to the south and west of the Plateau, probably caused by the Pacific plate's subduction, concentrated weight on the Earth's crust. The formation and subsequent disappearance of large lakes on the Plateau itself was a local influence. And finally, the great ice age of the Pleistocene Epoch had an isostatic effect when most of the North American

Transcontinental upwarp

There is a gentle "upwarp" of the North American continent from the Mississippi Delta to Grand Wash Cliffs, which are on the western boundary of the Colorado Plateau. This section shows the extent of this uplift (with the vertical scale greatly exaggerated). The horizontal line represents sea level.

ONE INCH = 15,000 FEET (VERTICAL)
100 MILES (HORIZONTAL)

LLANO ESTACADO

MISSISSIPPI BASIN

DELTA OF MISSISSIPPI

GULF OF MEXICO

continent, with the exception of its southern and southwestern areas, was covered by ice thousands of feet thick and this may have exerted some influence.

While the geological events that contributed to the uplift of the Colorado Plateau are complex and their interpretation contentious, there is neither doubt nor contention about the main force which carved the canyons of the Colorado River. It was water, the prime agent of erosion on this planet. There are few things more benign and beautiful than a clear stream meandering through a wooded dell on a brilliant autumn day. But there are few natural forces more destructive than a flash flood roaring through a steep narrow canyon after a heavy summer thunderstorm. Landscape changes in temperate climes are extremely slow but consistent; lush vegetation binds the soil, reducing the rate of erosion. Landscape changes in arid climates are inconsistent, sudden, and often devastating; vegetation is sparse, soil almost nonexistent, and the ground friable, or powdery. Grand Canyon and its brethren are products of an arid climate, and behind their beauty lies a savage force.

River water can carry solids in increasing proportion to its speed of flow. A river flowing at ten miles an hour can carry a million times the weight of material carried by a river moving at one mile an hour. At twelve and a half miles per hour the comparative carrying power increases to four million; and at fifteen miles an hour the force becomes a ravaging eleven million. River pebbles which tremble on the brink of movement in a river traveling slowly are replaced by heavily sedimented water tumbling house-sized rocks in a river swollen by torrential rain (see panorama at right). The action of such a river in flood is to grind and wear the riverbed, and if that bed is rising at the rate of one thousand feet per million years, it will cut downward at that rate. The determining factor in a river cutting a canyon is the angle of inclination. The steeper the angle, the quicker the erosion.

The original theory about the formation of Grand Canyon was proposed in 1875 by John Wesley Powell, the explorer-geologist who was the first to travel through the Canyon, in 1869. He suggested that the present Colorado River, fed by the melting snows of the mountains of the Wind River and Front ranges to the far north and northeast, and bolstered by the accumulation of storm water and sediment eroded from desert areas between the mountains and the Canyon, simply cut its way downward because the southern province of the Plateau began to rise.

But as the years went by, other more complex theories developed until the mind of an interested observer boggles at their intricacy. Common sense came to the rescue when in the summer of 1964 a symposium of twenty distinguished geologists, geophysicists, and paleontologists who had participated in practical work in the region, was held in Flagstaff at the Museum of Northern Arizona. Their objective was to review all the known facts about the evolution of the Colorado River and to hypothesize from them about the formation of Grand Canyon. The deliberations of the two-week-long meeting were presided over by Dr. Edwin D. McKee, the scientific father-figure of Grand Canyon. The series of events which were pieced together during the symposium was rather more complex than Powell's straightforward proposition.

At the end of the Cretaceous Era, the general area of northern Arizona slowly began to uplift and this continued until two distinct drainage systems had evolved in the region, the Hualapai and the Ancestral Colorado. The systems were separated by the

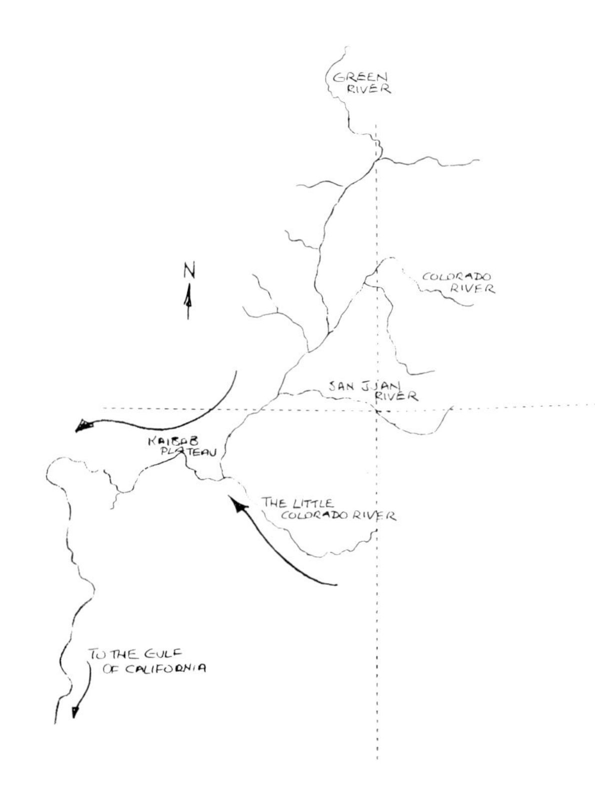

The Grand Falls of the Little Colorado

The spring runoff makes this one of the great spectacles of the Painted Desert. The river water is almost liquid mud, and falls over a hundred feet.
The river has eroded away thousands of cubic feet of the lava which once almost filled the canyon shown in this panorama. The whole scene is reminiscent of the Colorado River as it once cut Grand Canyon into the uplifting Kaibab Plateau. The heavily silted river water tears viciously at the Kaibab Limestone and the spray from the falls coats nearby rocks and cliffs with slippery red mud.
6. F, G-5.

View from Lipan Point, on the South Rim, near Desert View

Both the panorama and the diagram above show the curved uplift of the Kaibab Plateau at right in relation to the downward incline of the South Rim at left. The diagram is a section through the panorama and well beyond it to the southeast (left) and northwest (right).

The features have been exaggerated and simplified to clarify the step-like formations which relate Grand Canyon to Zion and then Bryce. The San Francisco Peaks are about 65 miles southeast of Grand Canyon; Bryce Canyon is some 110 miles to the right. 7. F-4.

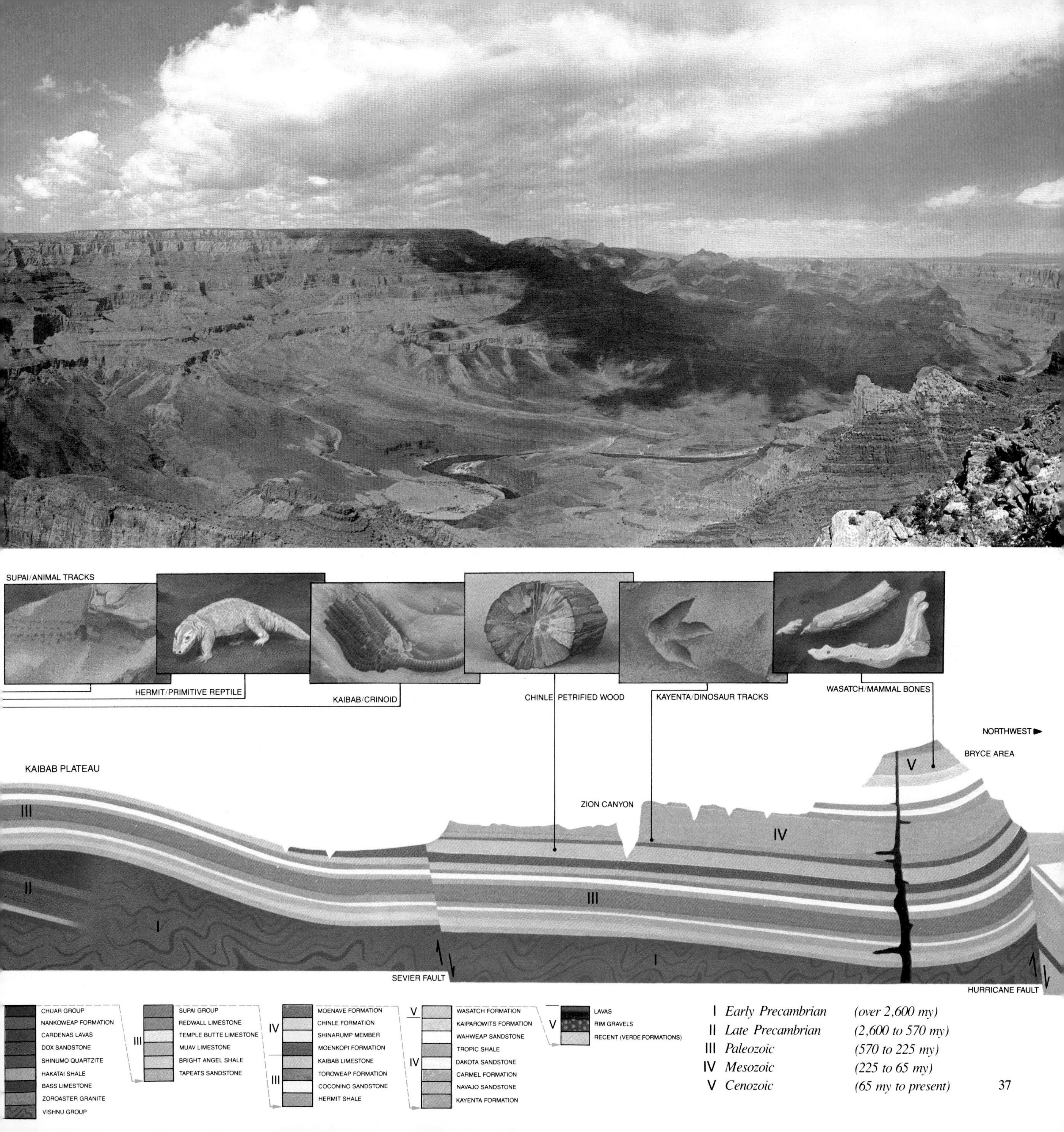
SUPAI/ANIMAL TRACKS
HERMIT/PRIMITIVE REPTILE
KAIBAB/CRINOID
CHINLE PETRIFIED WOOD
KAYENTA/DINOSAUR TRACKS
WASATCH/MAMMAL BONES
NORTHWEST
BRYCE AREA
KAIBAB PLATEAU
ZION CANYON
III
II
I
IV
V
III
I
SEVIER FAULT
HURRICANE FAULT
CHUAR GROUP
NANKOWEAP FORMATION
CARDENAS LAVAS
DOX SANDSTONE
SHINUMO QUARTZITE
HAKATAI SHALE
BASS LIMESTONE
ZOROASTER GRANITE
VISHNU GROUP
III
SUPAI GROUP
REDWALL LIMESTONE
TEMPLE BUTTE LIMESTONE
MUAV LIMESTONE
BRIGHT ANGEL SHALE
TAPEATS SANDSTONE
IV
MOENAVE FORMATION
CHINLE FORMATION
SHINARUMP MEMBER
MOENKOPI FORMATION
III
KAIBAB LIMESTONE
TOROWEAP FORMATION
COCONINO SANDSTONE
HERMIT SHALE
V
WASATCH FORMATION
IV
KAIPAROWITS FORMATION
WAHWEAP SANDSTONE
TROPIC SHALE
DAKOTA SANDSTONE
CARMEL FORMATION
NAVAJO SANDSTONE
KAYENTA FORMATION
V
LAVAS
RIM GRAVELS
RECENT (VERDE FORMATIONS)
I *Early Precambrian* *(over 2,600 my)*
II *Late Precambrian* *(2,600 to 570 my)*
III *Paleozoic* *(570 to 225 my)*
IV *Mesozoic* *(225 to 65 my)*
V *Cenozoic* *(65 my to present)*

The canyon trinity

This illustration relates the geology, topography, and geography of Grand, Zion, and Bryce canyons. The heights have been greatly exaggerated to allow features to be recognizable. The front section of each block runs from west to east. The Grand Canyon block cuts through Desert View (see panorama pages 36–37) and runs due west in a seventy-mile square. The Zion block cuts due west through Mt. Kinesava (see panorama pages 30–31) at the head of Zion Canyon and represents an area thirty miles square. The Bryce block, also thirty miles square, shows the relationship between the Paunsaugunt Plateau and the rim of the Bryce escarpment. The bottom right corner of the block cuts across the Paria Amphitheater and due north through Table Cliffs (see panorama pages 46–47). The Bryce area is drained by the Paria River, which joins the Colorado River at Lees Ferry. The Virgin River, which drains Zion Canyon, joins the Colorado below Grand Wash Cliffs in what is now Lake Mead, at the western end of the system. The Colorado River itself turns from due south to a westerly direction and winds its way for 200 miles below Desert View to Grand Wash. Kanab, Havasu, and Bright Angel canyons are the principal tributary canyons. The road system is schematic, intended only to show the relationship between the canyons. The geology has also been simplified.

ZION

HARMONY MOUNTAINS
ASH CK. RES.
KOLOB
BLACK RIDGE
91
LA VERKIN CREEK
THE BISHOPRIC
ANGELS LANDING
ZION CANYON
COUGAR MT.
HURRICANE MESA
W. TEMPLE
E. TEMPLE
NORTH CREEK
CRATER HILL
ST. GEORGE
15

NAVAJO SANDSTONE
KAYENTA MUDSTONE
MOENAVE SANDSTONE
MOENAVE MUDSTONE
ALLUVIUM AND SLIDE DEPOSITS

The Virgin River undercuts and collapses the steep walls of Navajo Sandstone of Zion Canyon, above.

GRAMA CANYON
HACK CANYON
JUMP UP PT.
JUMP UP CANYON

GRAND CANYON
AND THE
KAIBAB PLATEAU

KANAB PLATEAU
KANAB CREEK
FISHTAIL
INDIAN HOLLOW
GT. THUMB MESA
FOSSIL BAY
POWELL PLATEAU
TOBAR TERRACE
ESPLANADE
LEE CANYON
TOPOCOBA HILLTOP
AZTEC AMPHIT.
DOX CASTLE
MONADNOCK AMPHITHEATER
WALLA VALLEY
PT. SUBLIME
CONFUCIUS TEMPLE
SHIVA
COYOTE CANYON
LONG MESA
HUALAPAI CANYON
HAVASU CANYON
COCONINO PLATEAU
DIANA TEMPLE
TOWER OF RA
OSIRIS
TOWER OF SET
CRYSTAL RAPIDS
DANA BUTTE

BRYCE
- WASATCH FORMATION
- KAIPAROWITS FORMATION
- STRAIGHT CLIFFS, WAHWEAP
- TROPIC FORMATION
- DAKOTA SANDSTONE

ZION
- CARMEL FORMATION
- NAVAJO SANDSTONE
- KAYENTA FORMATION
- MOENAVE FORMATION
- CHINLE FORMATION
- SHINARUMP MEMBER
- MOENKOPI FORMATION

GRAND CANYON
- KAIBAB LIMESTONE
- TOROWEAP FORMATION
- COCONINO SANDSTONE
- HERMIT SHALE
- SUPAI GROUP
- REDWALL LIMESTONE
- TEMPLE BUTTE LIMESTONE
- MUAV LIMESTONE
- BRIGHT ANGEL SHALE
- TAPEATS SANDSTONE
- CHUAR GROUP
- CARDENAS GROUP
- DOX SANDSTONE
- SHINUMO QUARTZITE
- HAKATAI SHALE
- BASS LIMESTONE
- VISHNU GROUP

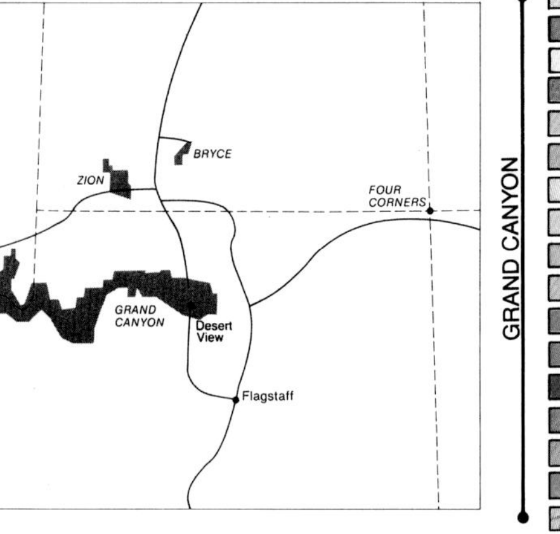

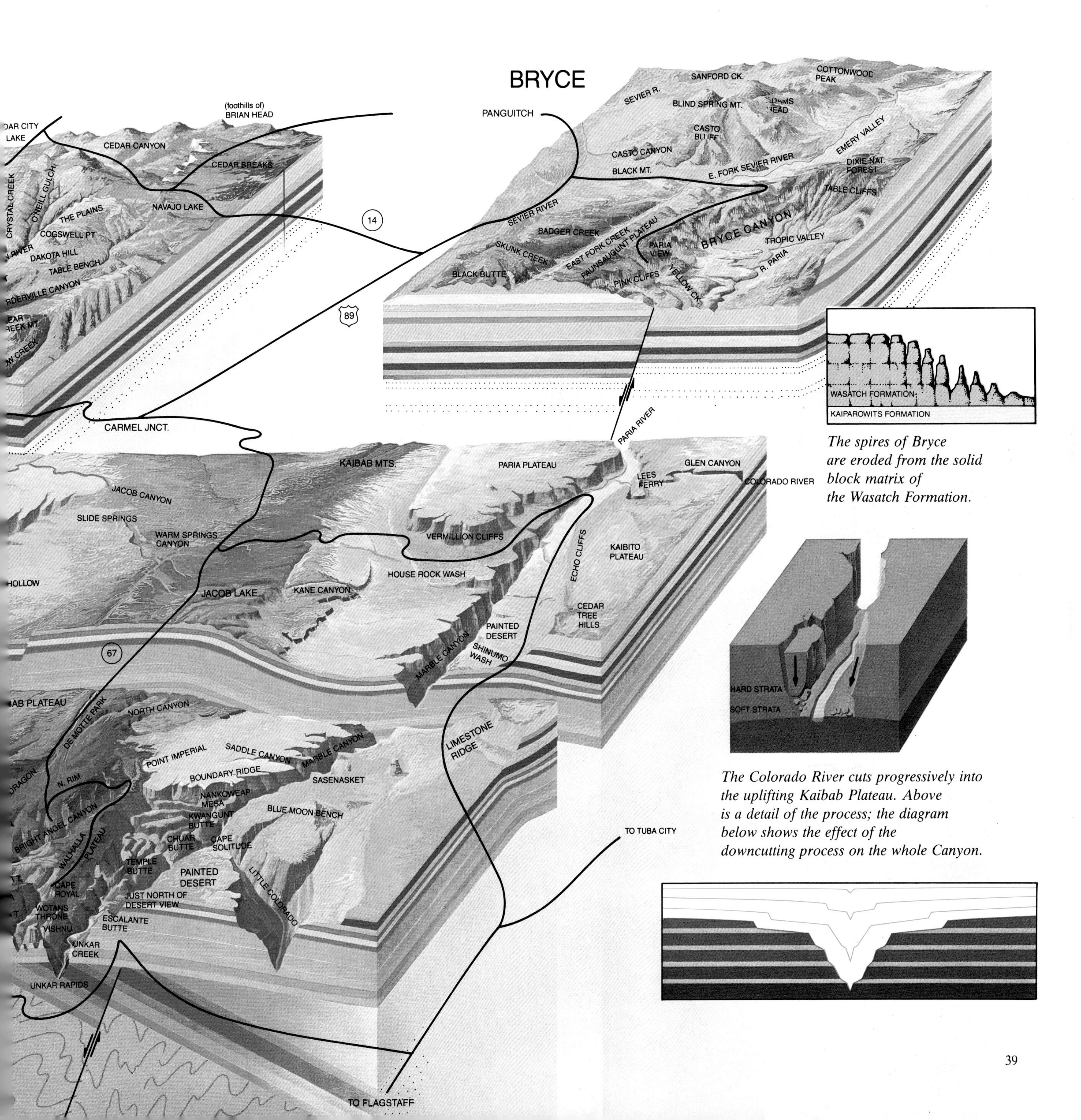

The spires of Bryce are eroded from the solid block matrix of the Wasatch Formation.

The Colorado River cuts progressively into the uplifting Kaibab Plateau. Above is a detail of the process; the diagram below shows the effect of the downcutting process on the whole Canyon.

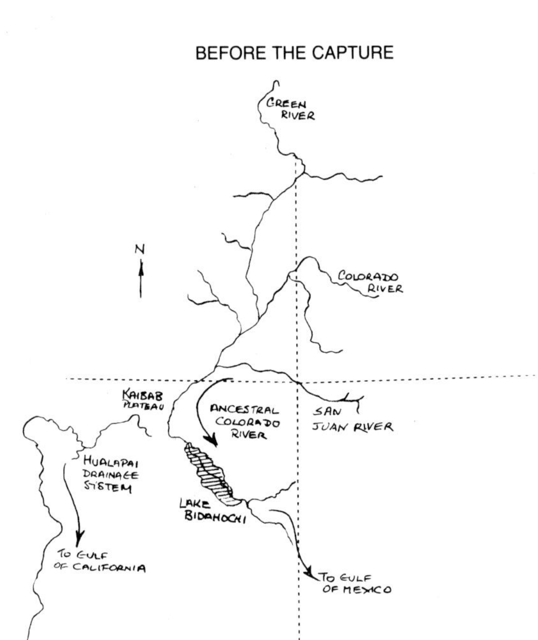

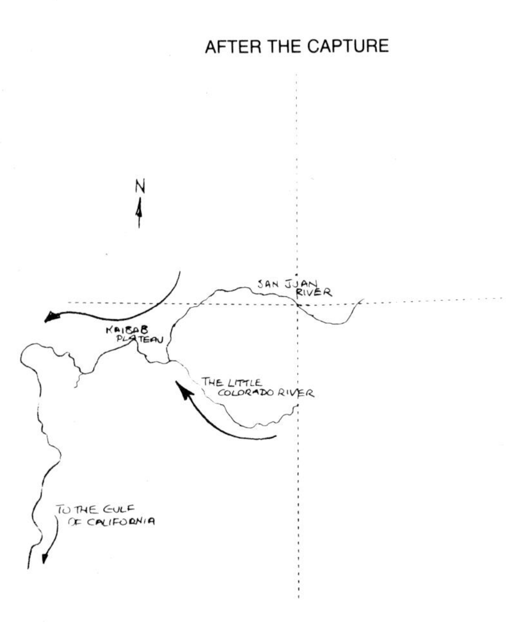

Plateau which at that time was only a few thousand feet above sea level. There was no Grand Canyon to join the two areas as there is today.

As the uplift of the Kaibab Plateau increased, the headwaters of the Hualapai drainage system cut back into the Plateau, thus establishing the pattern of flow of the future Colorado River through the western end of what is now Grand Canyon. Meanwhile, the Ancestral Colorado River, following the course of the present-day Little Colorado, but flowing in the opposite direction, formed a great lake called Lake Bidahochi. It is believed that the lake drained into the Rio Grande River system to the southeast and subsequently into the Gulf of Mexico.

As a result of the general process of uplift, the bed of Lake Bidahochi began to tilt. The Ancestral Colorado kept pouring into Bidahochi, while the headwaters of the Hualapai Drainage on the western side continued to cut back into the Kaibab Plateau. Then one momentous day, the last remaining rocks that separated the two drainage systems were gone. Eventually the contents of Lake Bidahochi emptied into the breach. The Little Colorado River reversed its direction of flow and Grand Canyon as we know it today began to form. The whole episode from the establishment of a dual drainage system to the "capture" of the Ancestral Colorado River was completed within a period of about five million years, commencing about ten million years ago.

In 1972, Dr. McKee and his son, Edwin H. McKee of the U.S. Geological Survey, a geologist specializing in tectonics and expert at radiometric dating, published a scientific paper that appears to clinch the conclusions of the 1964 Symposium. During a number of field trips along the southern margin of the Colorado Plateau, the McKees had found ancient streambeds, some well over a hundred feet deep, that had once drained in the opposite direction to the present slope. The size and rounded shape of the pebbles, cobbles, and boulders that the beds contained bore testimony to the force of the ancient torrents that had carved the beds, and their composition indicated a stream source from the south. What was crucial about the McKees' discovery was that while the streams were active, their beds had been filled by lava deposited by local volcanic eruption. This event established both direction and time of the stream activity—for lava can be accurately dated, and the direction of its flow is obvious to the observer. The lava had filled these gullies about ten million years ago. Similar lava flows on top of other nearby streambeds that flowed in the opposite direction when they were active were dated at about five million years. It followed that the direction of water flow in the locality of Lake Bidahochi had indeed completely reversed. Like the finding of *Lystrosaurus* in the Antarctic, this was the "incontrovertible proof" that geologists needed to establish that the period of greatest uplift of the southern Colorado Plateau, which resulted in the cutting of Grand Canyon, started about ten million years ago and was completed five million years later. Also that the direction of river flows had been reversed by uplift during this period.

There is a minority opinion about the formation of Grand Canyon that disagrees with the river-capture theory and at least partly goes along with Major Powell. The counter theory suggests that before the uplift of the Colorado Plateau started, the Ancestral Colorado did not flow anywhere near the Southern Province. It suggests that the Little Colorado River has always flowed in its present direction, and that in concert with the San Juan River from the northeast, it formed the true "ancestral" river. The Green and Colorado rivers were diverted from the north to add their force to that of the Little Colorado to form Grand Canyon during the period of greatest uplift.

One of the best places for any visitor to Grand Canyon to contemplate these theories and to visualize the processes of erosion involved is from Lipan Point on the South Rim of Grand Canyon, a few miles inside the East Entrance of the National Park. The view from Lipan is one of the most spectacular of the Canyon and it is also one of the most complete in the geological sense. According to the river-capture theory, somewhere in the immediate vicinity below the overlook, there took place quietly, almost imperceptibly at first, one of the greatest natural geographical changes of all time when the Hualapai drainage system stole the Ancestral Colorado River.

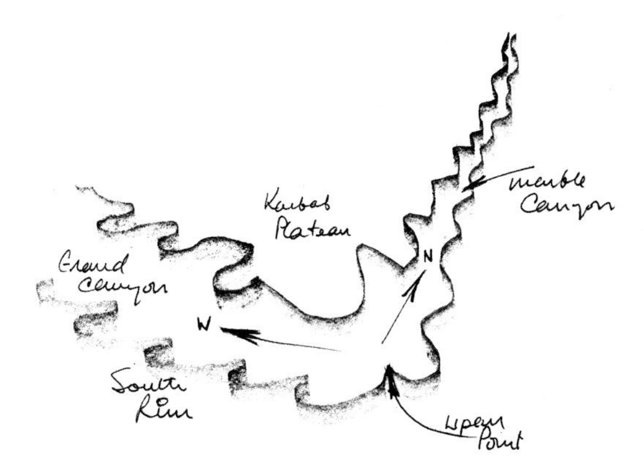

Lipan Point is a particularly good place from which to study the architecture of Grand Canyon, for it is at this point that the Colorado River turns ninety degrees from north/south to east/west. Consequently one looks due north up the Canyon and due west down the center of the Canyon, from this view point. Not only is the Canyon a vertical mile deep along much of its length, but it is at times more than ten miles wide. Between its walls are mountain-sized buttes and deeply incised side canyons; a world within a world. The profile of Grand Canyon as observed looking west from Lipan (see illustration and panorama pages 36–37) is the result of four elements working in combination: the uplift of the Colorado Plateau; the downcutting of the Colorado River; the uniformity of underlying rock formations; and the varying degree of hardness of rock types from very hard to very soft.

When a river cuts uniformly through hard rock as a consequence of uplift, a canyon with near-vertical walls will form. If uplift continues, the river will eventually meet underlying softer rock. The soft rock will erode more quickly than the hard. At first it forms steep sloping scree and when this is swept or worn away, undercutting follows. An overhang develops. When the overhanging rock can no longer support its own weight, it will fracture along joints or faults and fall. The debris from rockfall might form a rapid in the river below. The full force of sedimented river water will eventually erode the rocks that have formed the rapid, from boulder-size to cobble, then to pebble, and to granule size. The rapid will disappear and the river will cut its way into the next rock formation. Grand Canyon, consisting of twenty separate rock formation layers, was formed by this undercutting and widening process (see illustration page 39).

While uplift and downcutting by water erosion are the principal forces in Grand Canyon, there are other potent if more subtle forms of erosion constantly at work. Chemical changes in some rock will weaken its structure so that it will eventually crumble. The freezing and thawing processes of winter and spring, and the hot days and cool nights of summer, will cause weakening along natural planes of cleavage. A sudden wind squall can topple a column of rock that is split from a cliff face and ready to fall. Torrential summer rainstorms produce flash floods that can trigger a chain reaction rockfall; and an earthquake can cause the most immediate change of all.

The shape and character of eastern Grand Canyon are best appreciated from Lipan Point looking west, while its history can be reviewed at a glance by looking north. So important are these prospects of the Canyon to the understanding of it and of the geological history of the Earth, that to get the best possible view of the scene I secured permission to land a helicopter on Escalante Butte, about two thousand feet below Lipan but still three thousand feet above the floor of the Canyon.

After reconnaissance in a light airplane, and a closer look at alternative sites by helicopter, I landed just after dawn one day on the lower part of the Supai Group on Escalante Butte. Gear was hurriedly unloaded in the blast of the roaring rotor blades. I turned my back on the bedlam as the pilot opened the throttle. In seconds the contraption had gone, instantly vaporized in the vast stillness of the Canyon. Perhaps man had never trodden here before? The wonder of the place. I stood transfixed by the view for what seemed minutes before I could galvanize myself into action.

The platform toward the extreme end of Escalante narrows to a promontory of crumbling Redwall a few feet wide and fifty yards long. On either side of this stairway in the sky I had the illusion of an abyss of three thousand feet with perpendicular walls on either side. It was not at all comforting to remind myself that the Redwall Formation is at worst only six hundred feet thick and if I slipped that was the maximum distance I could fall. My objective was to reach the dome-like structure at the end of the narrow section separated by a dip of a hundred feet or so, and then climb up an unpleasantly steep gully of loose rock of equal length to the top. It took careful concentration to reach the summit, only to find that the summit of the dome was so limited in area that I had to build a small platform from loosc rock for thc camcra tripod. Thcsc cngincering efforts didn't please a large menacing tarantula wasp which had a nest nearby. I was once told by a naturalist in answer to the question, "Do tarantula wasps sting humans?" that "They do have the apparatus," so I kept a weather eye open for squalls. After a few aggressive swoops I was left in peace.

The sight around me was not of this world. There to the north lay a section through part of continental Pangea, sharply etched in the low-angled early light of day. Below it lay all that remains in this region of a previous primordial continent, Pangea's unnamed Precambrian predecessor. The central feature of the northern aspect of the

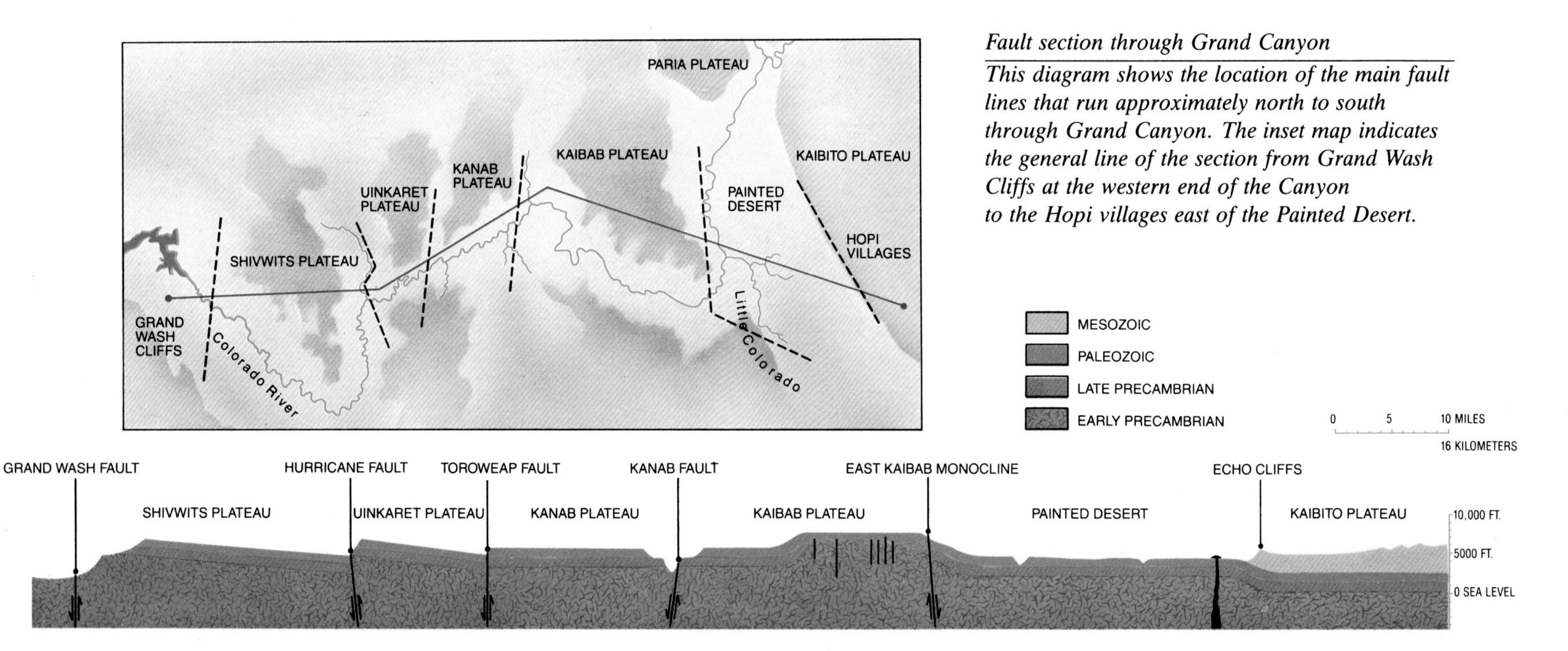

Fault section through Grand Canyon

This diagram shows the location of the main fault lines that run approximately north to south through Grand Canyon. The inset map indicates the general line of the section from Grand Wash Cliffs at the western end of the Canyon to the Hopi villages east of the Painted Desert.

view from my perch was Unkar Rapid in the bottom foreground (see panorama pages 15–18) and the clearly defined strata of the Precambrian formations above it. The angle at which the whole series is set clearly identifies its extent. Each of these sloping layers represents the compressed remnants of stages in the build-up of a Precambrian continent. The youngest stratum of this substantial remnant is certainly more than six hundred million years old, and the oldest perhaps twelve hundred million. These sediments of ancient erosional cycles contain fossilized algae, one of the earliest forms of life on Earth. The search for clues to early forms of life goes on in the region. Here lies the record of an embryonic world.

Where the tilted Precambrian formations end, some of the underlying rocks of the pre-Pangea continent begin. The oldest of these rocks are approximately five hundred million years old and form the Tonto Platform, which is a dominant feature of much of the eastern part of Grand Canyon below the popular South Rim. In these rocks and the immediately overlying Temple Butte and Redwall Formations, fossils of a wide span of early sea life are found. Then follow the bright terra-cotta-colored Supai Group of the Pennsylvanian and Permian periods which commenced about three hundred and twenty million years ago. These contain a multitude of early fossil plant forms including seed ferns and early amphibians not unlike *Lystrosaurus*. Rocks of the Supai Group dominated the landscape of this region for forty million years and bridged two periods, the Mississippian and the Permian. The topmost formation on both North and South Rims is also Permian—the Kaibab Limestone was a shallow seabed at the time that continental Pangea was almost fully formed (see illustration page 23).

Whether you see it for yourself from Lipan Point or just share the view from Escalante Butte with me, you are looking at an exciting scene from the geologic record. This is one of the best places on Earth to view a substantial section of part of the underlying rocks of the primeval continent of Pangea. The rocks you see were formed in a completely different geographical location, certainly hundreds and perhaps a

Types of faults

A fault is a fracture or fracture zone created when a rock formation is subjected to tension, compression, or shearing. The faults in the Kaibab Plateau region are mainly caused by the relief of the tension produced by uplift. Some of the faults have caused enough displacement of individual rock formations that continuity between them has been broken. Similar vertical displacement of rock formations also occurs in monoclines in which horizontal strata are down-folded steeply before returning to the horizontal. The steep strata are often cut by a fault, and monoclines themselves are frequently sited over fractures in the older basement. The east Kaibab monocline is an example of this. It can be seen in the section on the page opposite and in the panorama on pages 116–117.

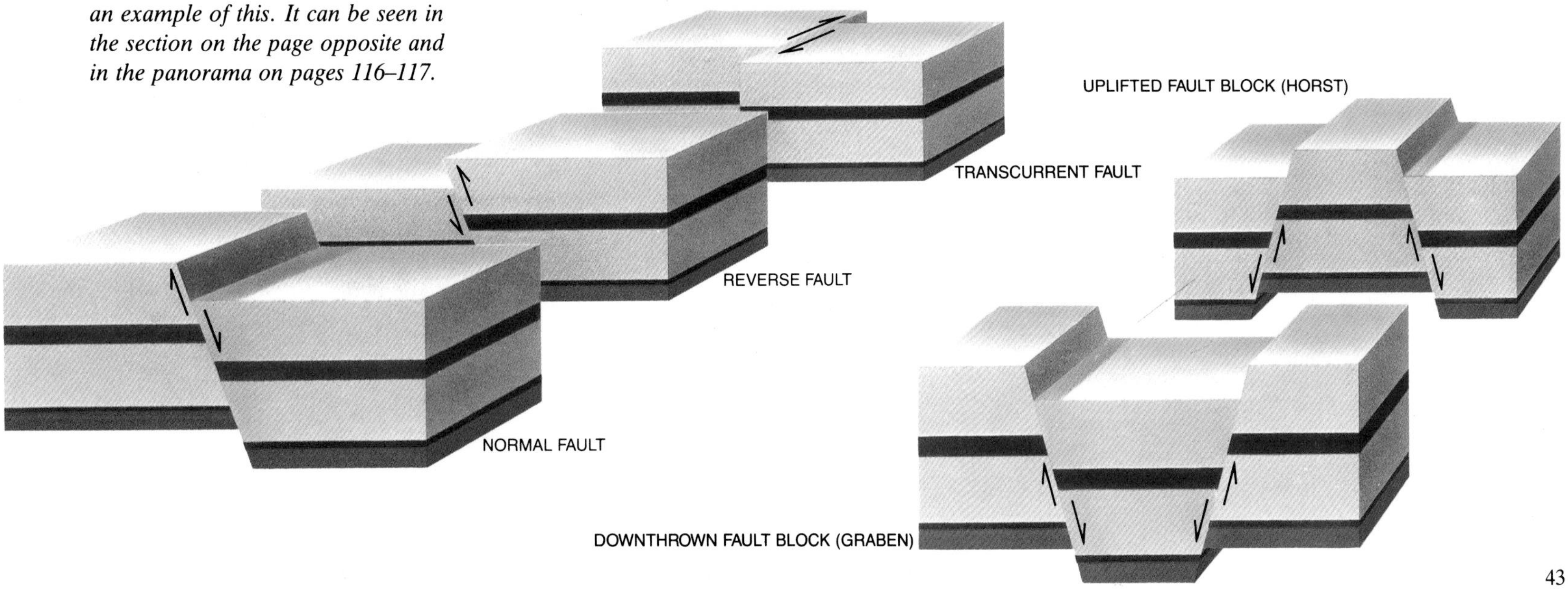

UPLIFTED FAULT BLOCK (HORST)
TRANSCURRENT FAULT
REVERSE FAULT
NORMAL FAULT
DOWNTHROWN FAULT BLOCK (GRABEN)

thousand miles or more from here. Elsewhere in the world where such sections have been exposed, underlying rock formations are often deformed by tectonic movement. Uniform layers of comparatively undisturbed rock are a prime ingredient for the making of canyons of the character of Grand Canyon.

Uplift causes tension and compression which is relieved by faulting (see illustration page 43). A combination of uplift, faulting, surface erosion, and river cutting has resulted in massive stripping of the landscape. For this combination of reasons the original formations above the rims of Grand Canyon, hundreds of square miles in extent and well over a mile in total thickness, have been removed during the Mesozoic and in this present Cenozoic Era. Grand Canyon skirts the southeastern and southern perimeter of the area of greatest uplift, the Kaibab Plateau. The present limits of the stripping process lie to the north, where a series of step-like escarpments are obvious to the visitor traveling from one canyon to the other.

The first of the series of remaining formations rising above the Kaibab are called the Vermilion Cliffs. After a geographical step back in a generally northward direction, another series of cliffs rises above them—the White Cliffs, into which the canyons of Zion are cut. A further step back and then the Gray Cliffs above Zion. A final step back and the Pink Cliffs from which the spires of Bryce are formed. Above all these lie three high plateaus, the Markagunt, Paunsaugunt, and Aquarious which crown the whole complex. Zion Canyon is cut into the Markagunt Plateau. Its crags soar from lush canyon floors to reach straight for the sky. Bryce Canyon is cut into the Paunsaugunt Plateau, and is an unreal world of fantasy and color, a confection of some of the youngest rocks on Earth (see illustration pages 38–39).

Both canyons were formed as Grand Canyon was formed: by the combined effect of uplift and water erosion. The Virgin River is the cutting force that shaped Zion. The headstreams of the Paria River carved Bryce Canyon. Both the Virgin and the Paria rivers are tributaries of the Colorado River. Because the strata that can be seen in these canyons once covered much of the region of Grand Canyon, erosional forms similar to Zion and Bryce may have once occurred nearer to Grand Canyon during the process of denudation of the surfaces above the Kaibab Limestone.

The dominant rock formation in Zion Canyon is hard, thick, but permeable, sandstone. The underlying formation is comparatively soft mudstone which is slippery when wet. The resultant interaction between the two rock types is almost predictable. The downcutting Virgin River runs across the bare rock of the Markagunt Plateau above, cutting a deep narrow channel called The Narrows, until it reaches the softer rock underneath. The undercut rock then shears under its own weight and falls. But the process at Zion is further complicated because water retained in the sandstone will soften the mudstone and cause a weakening of the foundation on which the thick layer of sandstone rests. The mudstone will provide support for a while, but eventually a considerable section of the sandstone cliff face will slump. The overall effect is widening and deepening.

Zion Canyon becomes more and more confined as one progresses up the Virgin River to The Narrows at its northern limit. In some places the vertical walls of The Narrows are more than two thousand feet high but only twenty to fifty feet apart. For most of the year the river flows demurely down a long incline from the plateau above into The Narrows between walls gradually increasing in height. In spring the melting snow, and in summer heavy rainstorms, transform the Virgin into a raging virago. The depth of the river can rise from its normal foot or so to as much as twenty or thirty feet. The destructive force as the torrent roars down that narrow defile can be imagined. The rock cliffs nearest to the exit from the narrow section eventually are undercut and collapse. When this event occurs, rock debris blocks the exit until it too is reduced to gravel and sand. Imperceptibly, Zion Canyon grows longer and wider (see panorama pages 30–31).

There are many places in the canyon where the evidence of water permeability of perpendicular sandstone cliffs can be seen. The best-known example is Weeping Rock, where a perpetual stream of crystal-clear water droplets drain from a reservoir in the rock mass above and fall from the edge of an overhang (see panorama page 44). In wintertime the droplets freeze and accumulate into gigantic icicles. The water in the rock remains above the freezing point except in the vicinity of the rock surface. In extremely cold weather, icicles with a girth of several feet and a height of twenty or thirty feet grow like stalactites until they break off under their own weight. The wet, soft, slippery rock beneath the overhang is called mudstone, and it is this that is subjected to the shock of falling icicles that weigh many tons. In time, either the overhang will collapse when it can no longer support its own weight, or the soft underlying mudstone will weaken until it offers no support and the sandstone cliff will slump. It is this kind of additional erosional process, together with the downcutting of the Virgin River flowing along the canyon floor, that has widened Zion Canyon beyond the mouth of The Narrows.

The Virgin River flows south out of Zion and then turns west passing through a small town called Rockville. South of the town stands an imposing array of lofty buttes that rise tier after tier from the desert floor. Looking back toward Zion from these buttes one can see a two-hundred-million-year-old accumulation of rock formations. The lowest of these formations formed the landscape of Pangea and the later ones the landscape of Laurasia (see illustration page 23). These formations accumulated as the North American continent drifted first northward then westward to its present geographic position (see panorama pages 30–31 and illustration page 23).

The Kaibab Limestone which forms both North and South rims of Grand Canyon also forms the lowest rock formation of the Rockville Vista. The youngest rocks of Grand Canyon are therefore the oldest rocks at Zion. Immediately above them are rocks that are the sediments of river estuaries that existed in early dinosaur times, the same rocks that form the Painted Desert. The whole character of Zion is, however, dominated by two formations totaling well over two thousand feet in thickness. The lowest stratum of Kayenta Formation, mudstone and siltstone, was formed from stream deposits, but the Navajo Sandstone was formed in a vast sand desert that was as extensive in this area during late Triassic and early Jurassic times as similar deserts in North and Southwest Africa are today. This landscape was, in fact, rather nearer the Equator when it was formed than either of these modern deserts and very likely was hotter.

The Navajo Sandstone formation of Zion is capped with three hundred feet of Carmel Limestone in which marine fossils are prolific. The Carmel is the youngest

Weeping Rock (opposite)

The vast sandstone formations of Zion act as a kind of reservoir. The porous rock allows rainwater and melting snow to percolate and drain through, as can be clearly seen here. This contributes to undercutting and eventual collapse of the cliffs. In winter the constant seepage freezes. Icicles grow to enormous size, draping down from the overhang. During a period of thaw these collapse and further weaken the foundations of the overhang. 8. D-3.

The Paria Amphitheater, Bryce Canyon (overleaf)

This panorama was taken in February at a time of record snowfall. The temperature in the shadow was well below zero Fahrenheit at an elevation of some 8,500 feet. Much of the Paria Amphitheater can be seen from this point (see illustration pages 38–39). The Table Cliffs, left background, were once part of the Paunsaugunt Plateau, but faulting caused that area to be elevated nearly 2,000 feet above the present level of the Plateau. Erosional forces can be seen at work here. Ice from the outer layers of soft rock is melting in the sun, although the air temperature is extremely low. Accumulations of small pieces of rock which have been eroded by the freeze-melt cycle can be seen in the discolored snow at the foot of the sunlit pinnacles. 9. C-4.

formation at Zion and forms the surface of the final step in the series that starts with the rim rocks of Grand Canyon and finishes with the youngest in the region, at Bryce Canyon, forty miles northeast of Zion. A steady uphill climb has to be negotiated to reach Bryce and during the course of this distance, rocks rise layer after layer on top of the Carmel Formation until, at an elevation of nine thousand feet, the undulating surface of the Paunsaugunt Plateau is reached. This, the top of the pile, is over a mile above the level of the Colorado River in Grand Canyon, one hundred miles due south.

Bryce is not a canyon but an escarpment. It is the scalloped eastern edge of the Paunsaugunt Plateau, bristling with many thousand multicolored castellated spires. The spires vary in height up to several hundred feet and rise from steeply inclined ridges fanning downward to the Paria Amphitheater below (see panorama pages 46–47 and 62–63).

The Paunsaugunt Plateau slopes westward from the escarpment. None of the streams that form the Sevier River that originates on the plateau contribute to the erosion of Bryce in any way. The rim of the Canyon is a watershed between an east-facing slope and a western slope. The western slope is so gradual that the Sevier River, which has run its course since Bryce Canyon began to form, still occupies a shallow bed that winds its way off the plateau and then northward to play no further part in this story. The eastern slope is a different matter, because it is very steep and the cutting power of rainwater from torrential summer storms is therefore considerable.

The nature of the rocks at Bryce Canyon determines the steepness of its slopes, the form of its incredible multifarious spires, and its fabulous colors. The Wasatch Limestone, its principal constituent, is young—only forty to fifty million years old—and appears to be more of a loose crystal aggregate than a rock form. Because of the high elevation, nearly ten thousand feet, the water that penetrates the surfaces of the rock freezes on winter nights and thaws in the sun on most days. Add the other ingredients of erosion, melting snow in springtime and heavy rain in summer, and it is easy to understand both the steep incline and the deep channels between ridges. Water will always follow the shortest and steepest route, so the same gullies will continue indefinitely to cut down and back into the escarpment as long as something of the Paunsaugunt Plateau remains to be cut. It is interesting to conjecture what will happen when the process of cutback reaches the Sevier River. When the breach occurs, another river-capture episode will have occurred with perhaps considerable effect on the whole Paria Amphitheater.

Those preposterous pink spires that seem to appear ready-formed from the mantle of the escarpment face are, in fact, initially the rectangular blocks of the solid-jointed formation of the Wasatch. As the loose escarpment surface is washed away, the cliff walls emerge and themselves become subject to the process of erosion. First, they are divided into separate vertical walls by the cutback action of the stream gullies which leach into joints in the rock formation. Then the "wall" erodes into separate spires (see illustration page 39). If the top of a particular length of wall is surmounted by a harder form of rock than the underlying material, it will be the latter and not the former that will erode first, causing a change in vertical profile. If the overlying hard material covers only a small area of a wall, then a freak shape, called a "hoodoo," will result. Eventually the softer rock will erode so that it can no longer support the weight of the rock now perched precariously on top of it. At this time the whole edifice will collapse. The varying heights and shapes of the myriad spires, walls, castles, cathedrals, ships, and courts are caused by variations in the degree of resistance of individual layers that make up the friable rock.

The rocks of Bryce Canyon were formed from lake sediments that contained considerable quantities of iron compounds intermixed with sands and gravels from freshwater streams. Before exposure, the limestones are pale pink and the sandstones and gravels are an off-white in color. But once exposed to weathering, the iron content becomes subject to chemical changes that produce the terra-cotta reds, the pinks, yellows, tans, and even delicate mauves. The distribution, density, and type of iron compounds differ, producing an astonishing spectrum of color that alters in intensity and hue according to the time of day. The cold light of dawn accents the blue end of the spectrum, and the warm glow of the evening sun exaggerates the red.

Below the Wasatch lie three other rock formations representing the Cretaceous, a period of sixty-five million years during which the age of dinosaurs came to an end and following which the age of mammals began. The eastward view from the rim of Bryce Canyon across the Paria Amphitheater (see panorama pages 46–47) is the counterpart of the views of the Rockville Vista (see panorama pages 30–31) and the Unkar Creek region of Grand Canyon (see panorama pages 15–18). They each portray particular stages in the story of a dynamic planet. Collectively they give us a glimpse into the history of the Earth, covering nearly seventeen-hundred million years. They provide a means of studying events on Earth during the assembly and the breakup of the primeval continent of Pangea. They help us to visualize what ancient landscapes looked like. They truly are corridors of time.

III

1,700,000,000 YEARS OF EARTH

The Painted Desert at first light

A "primordial landscape" dominated by the Chinle Formation of Late Triassic times, and consisting of shales and siltstones ranging in color from grays and near-whites to bands of yellow, pink, and terra-cotta. It was formed in a river environment, some 200 million years ago. 10. G-6.

The men who wanted to become the first to walk on the Moon sat around the pool in the half-light after dinner one evening and listened to Dr. Edwin D. McKee of the U.S. Geological Survey. The party had hiked down the Kaibab Trail that day to Phantom Ranch at the bottom of Grand Canyon. Their purpose was to learn some basic geology, sufficient to be able to report accurately on what they would see if they were fortunate enough to participate in one of the Apollo missions. Dr. McKee was using his renowned expertise of Grand Canyon to teach the astronauts the principles of stratigraphy—the method of classifying, relating, and interpreting rocks.

The Moon, which one likes to think was hanging large and inaccessible that night, is a fossilized world preserved in vacuum. It formed simultaneously with Earth four billion six hundred million years ago. Like Earth, it is a result of the condensation of stellar dust. It was bombarded by asteroids and meteorites, and perforated by volcanic eruptions. Magma from the Moon's then largely liquid subsurface flooded large areas of its surface to form lava seas. The Moon's gravity is one-sixth that of Earth's and is too weak to retain an atmosphere. There is now little or no movement of the Moon's crust, just an occasional tremor, a modest moonquake. The Moon has a very weak magnetic field, signifying that it does not have an iron-nickel core like the Earth, and is now little more than a solid ball of rock. It is one of the many stillborn children of the solar system.

What the astronauts would confirm during their future exploration of the Moon's surface is that it has changed little in comparison with the Earth's surface during the three billion years that have passed since the end of the asteroid storm. During this immense period, the principal form of erosion on the Moon has been cratering by meteorites, but even this process has been diminishing. There have apparently been few major meteoroid impacts for at least a billion years. There are, however, countless mini-craters on the Moon's surface, the result of the impact of hundreds of pebble-size meteorites.

Surface temperature fluctuates by five hundred degrees between lunar day and night. The combined effect of expansion and contraction, together with radiation from the sun, disintegrates Moon rocks, so that crater walls weaken and slump, debris degenerates into fragments and then into dust. In a few million years astronaut footprints will be erased by this process. What little happens on the Moon happens slowly. There are no tectonic plate movements and therefore no continents. Because there is neither water nor atmosphere, there are no rivers, lakes, or oceans. There is no climate, and therefore no sedimentation. In fact, all Moon rocks are igneous, the result of melting and cooling processes, and are therefore crystalline in form. A mile down from the South Rim at Phantom Ranch, the astronauts were enveloped by such igneous formations, the oldest rocks found in Grand Canyon.

And so it was during the early history of Earth. All rock forms were the products of fusion by heat and subsequent crystallization. Primitive Earth suffered the

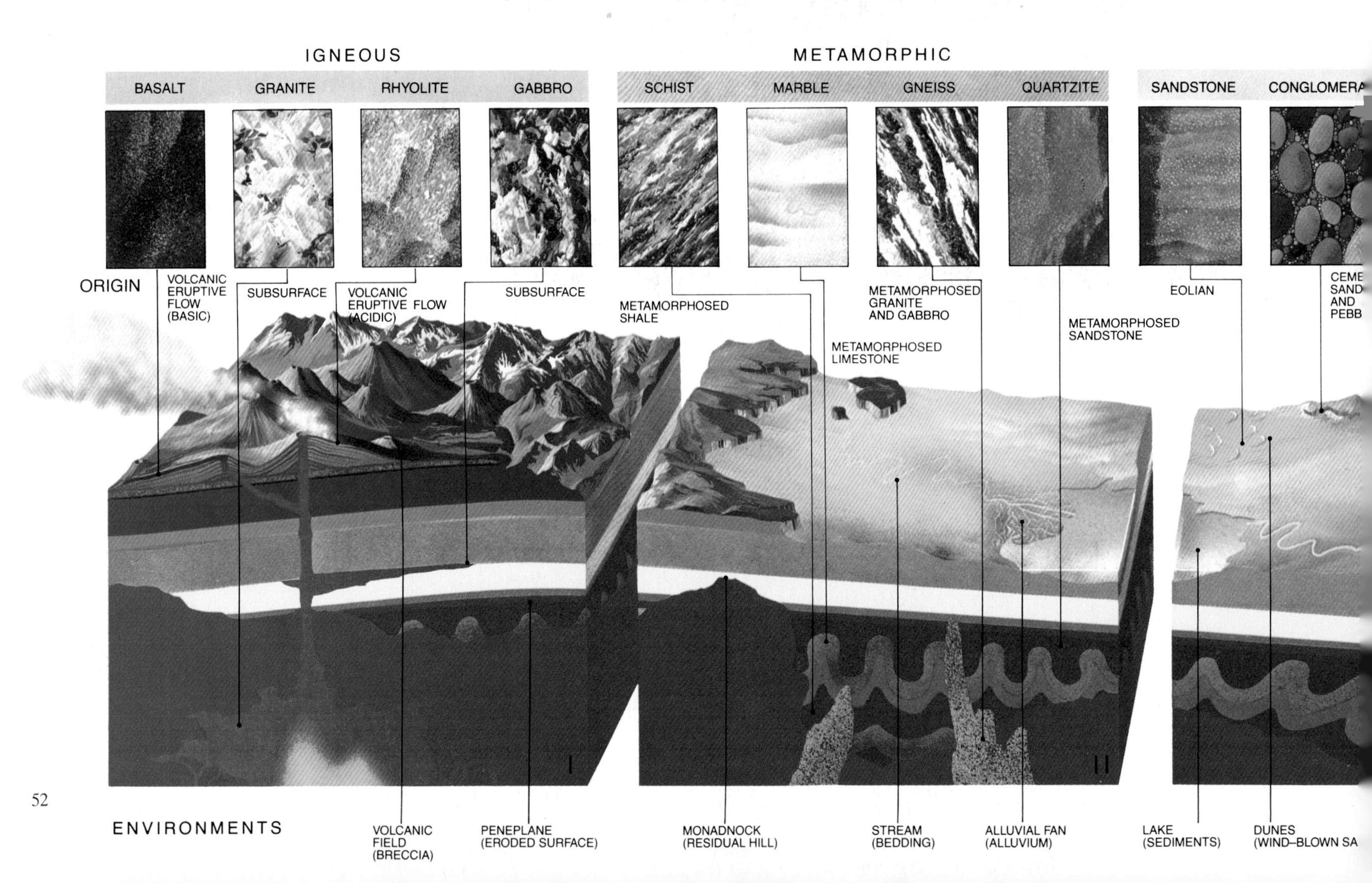

same bombardment of asteroids as the Moon and was riddled with similar violent volcanic action. But because of Earth's composition and mass, there was a very different consequence. The Earth is forty percent more dense than the Moon and four times its size, and these differences were crucial. Earth's volcanic emissions formed a primordial atmosphere and it was permanently retained by the Earth's gravity, which is a factor of its density and mass.

During its earliest phase of development, the Earth's surface consisted of semimolten basaltic materials on which less dense granitic matter floated. Beneath the hardening crust, heavy silicates and dense basalts merged in the mantle, while the heaviest materials of all gravitated to the center to become the Earth's extremely hot metallic core. By the time the core had formed, the crust had solidified, while the mantle remained plastic (see illustration page 22).

The crust formed early in the Earth's history. It consisted of basalt encrusted with granite, riddled with violently active volcanoes, bombarded by asteroids and meteorites, and subject to intense solar radiation. But the surface of the Earth was so well insulated from the mantle by the crust that only six hundred million years after its formation the surface temperature had cooled sufficiently to allow the high proportion of water vapor in the volcanic emissions to begin to condense. The subsequent story of planet Earth, so very different from that of its stillborn Moon and its fellow planets in

Formation of rocks

There are three kinds of rocks: igneous rocks, formed by the solidification of the Earth's magma; sedimentary rocks, which are composed of minerals, organic materials, and fragments of other rocks that have been suspended in water, air, or ice, and have then come to rest in the form of a bed; and metamorphic rocks, which can be either sedimentary or igneous rocks that have been reprocessed by the action of great heat. Sedimentary rocks cover more than two-thirds of the Earth's land surface (often only as a thin veneer). They contain clues to the conditions in which they were formed. Originally they may have been sand dunes blown by the wind (see panorama page 59). Or they may have formed on the bed of a stream or river, lake marsh, or delta, or, most frequently, on a seabed at the margin of a continent. They may contain fossils of plants or animals. The imaginary landscape below shows some of the many ways in which rocks and minerals form, and the environments in which they do so.

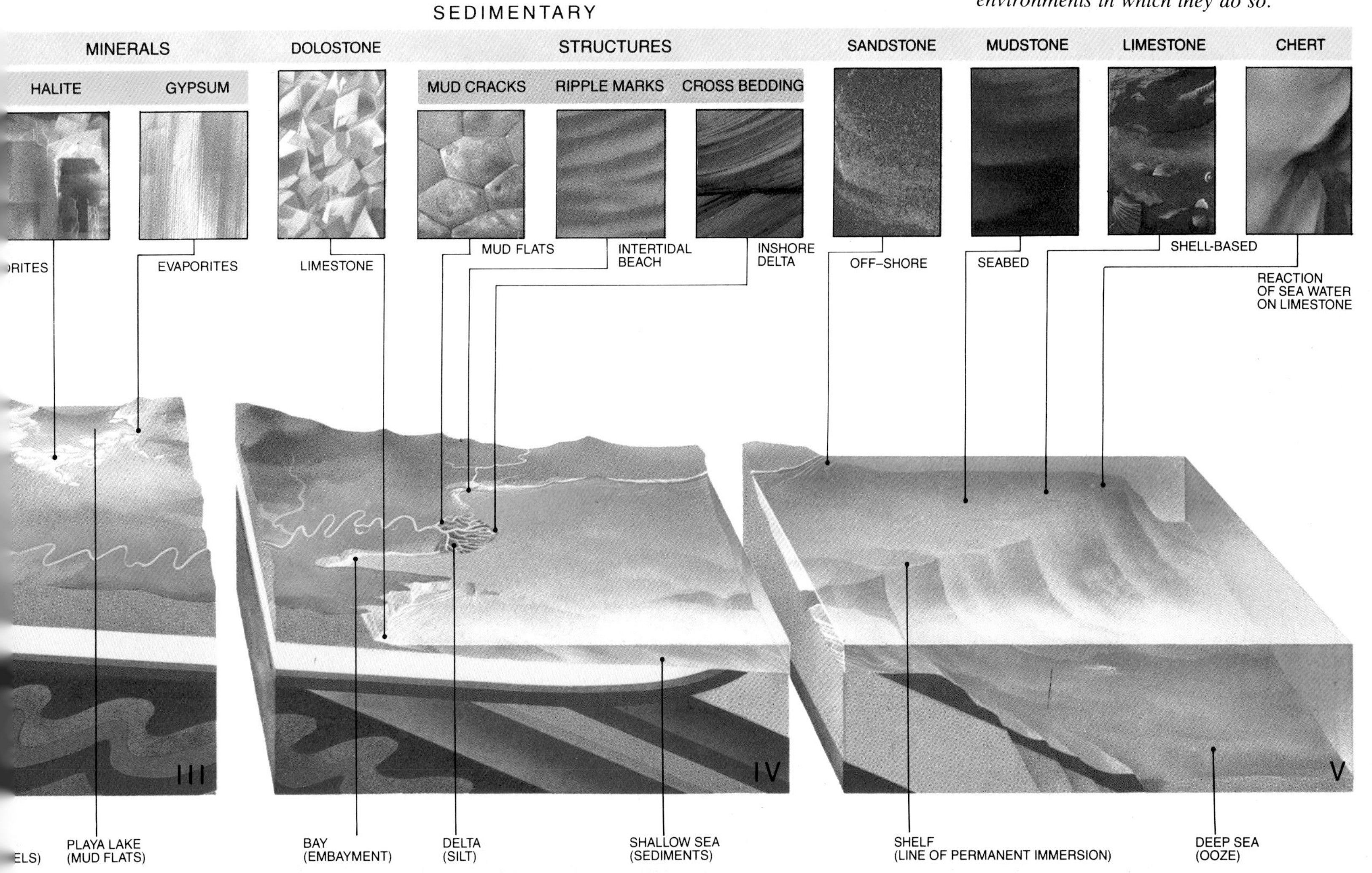

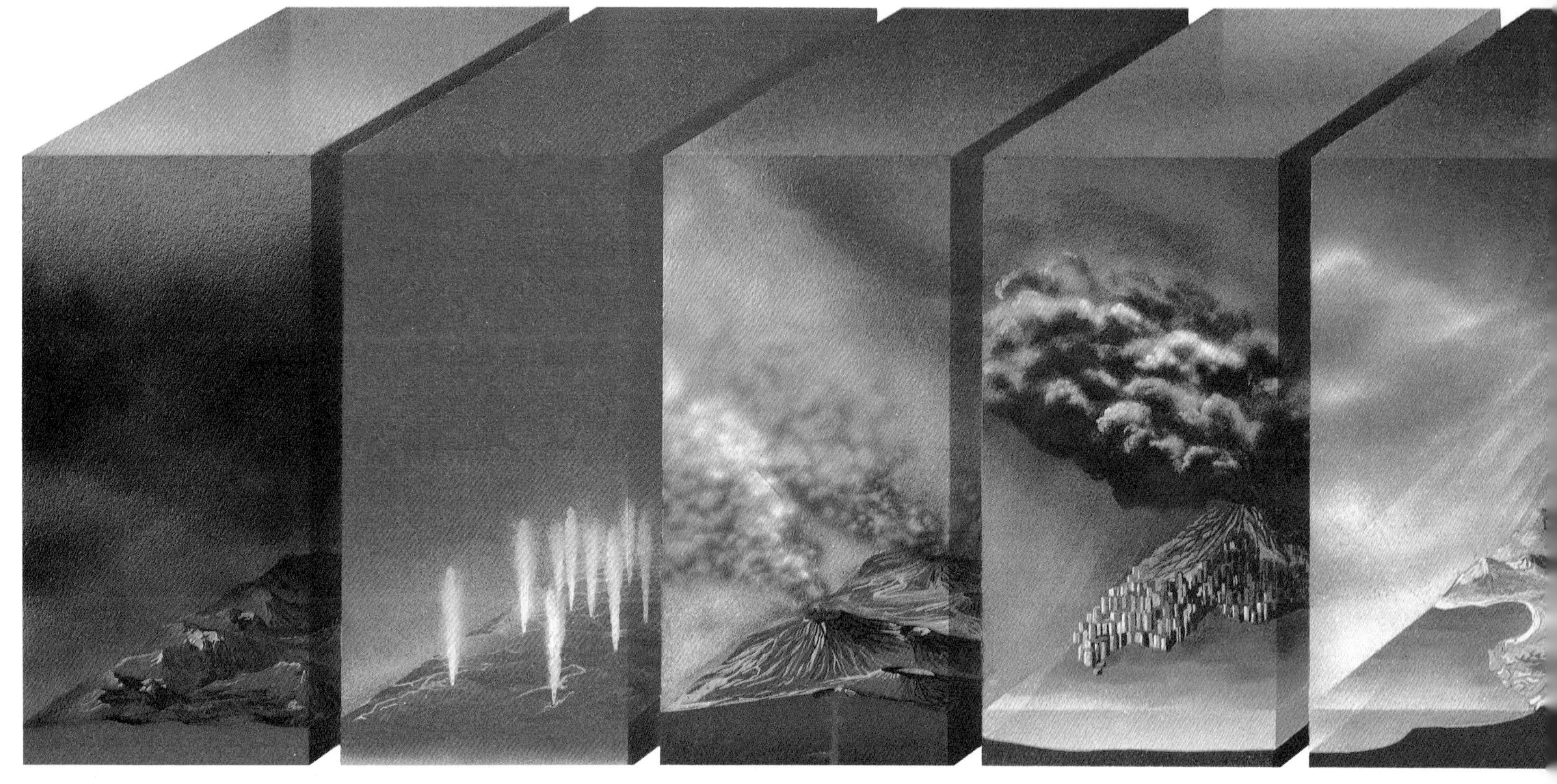

Primordial planet of cosmic gases and an iron-silicate magma.

Addition of gases from the molten surface. An atmosphere of hydrogen, methane ammonia, and water vapor.

Volcanic activity, terrestrial storms, and the sun's ultraviolet rays all acted upon the intensely hot planet.

Continuous condensation and cooling land masses heralded the building blocks of life—nucleic acids and other organic molecules.

Algae used sunlight to form free oxygen in the atmosphere, creating the ozone layer—an ultraviolet shield.

Evolution of the atmosphere

During the 4,600 million years of the Earth, the atmosphere has undergone many gradual changes in its stages of evolution. For most of this time the atmosphere was inhospitable to life as we know it today. In the first stage the primordial gases enveloped the planet but were lost to space—they were too light to be retained by Earth's gravity. There was then a long period of exhalation of gaseous material from the semi-molten surface, gradually replaced by volcanic emission. The oceans were formed by the precipitation of water vapor from these emissions, and the process of erosion began. As this progressed, the first simple sea plants formed which photosynthesized their sustenance and formed oxygen. In the early stages of oxygen production, capacity was taken up by other elements and substances. The first free oxygen began to accumulate when this process had been completed, and the atmosphere resulted.

the solar system, began with the very first raindrop that hissed and evaporated on the hot igneous rock on which it fell.

While theories about the very early history of Earth are speculative, current research suggests that there were three stages in the development of the atmosphere, and that these determined the environment of the world we live in and the sometimes very different environments we can see portrayed in the fossil record at Grand Canyon, Zion, and Bryce.

Earth was in a semisolid state during the first several hundred million years after it had formed. The light elements hydrogen and helium were abundantly present, and together with argon and neon formed a transitory atmosphere that escaped the Earth's gravitational pull. The heavy elements, nickel and iron, began to gravitate to the Earth's center to form a core while the surrounding temperature of the asthenosphere escalated as a by-product of both gravitational pressure and radioactive decay. On the surface, volcanic activity, which emanates from the asthenosphere, was rampant as the Earth's crust cooled. The light gaseous elements that had been lost to space were replaced by heavier volcanic gases which formed the primordial atmosphere. This atmosphere consisted mainly of water vapor in a proportion perhaps as high as eighty percent. The balance was made up from carbon compounds such as

Multicellular life-supporting atmosphere of oxygen, carbon dioxide, and nitrogen formed in late Precambrian times.

carbon monoxide and dioxide, hydrogen compounds of sulphur and chlorine, and nitrogen. Oxygen is a rather subservient element and readily combines with many other elements. It is also fickle and will easily transfer its affections from one element to another. For this reason, no free oxygen existed in primordial conditions because there were so many substances available with which it could combine.

In the second stage of the evolution of Earth's atmosphere, the predominant component was still water vapor, but the makeup of the other twenty percent changed significantly; carbon dioxide predominated, followed in order of magnitude by nitrogen and sulphur dioxide. There was still no free oxygen. Photosynthesis, the ultimate system of oxygen production, was yet to come in the third stage of atmospheric evolution. In fact, more than half the four-thousand-six-hundred-million-year history of Earth passed before the proportion of oxygen to other available elements in the atmosphere became significant and our present atmosphere began to take shape.

It has been estimated that there are one hundred and forty-five thousand billion billion metric tons of water contained in the seas, on the land, and in the atmosphere. Most of this incredibly large volume has been the consequence of emissions from volcanoes and fumaroles, water which must have been dissolved in the molten magma. The rest, thought to be a comparatively minor proportion of the whole, is the by-product of chemical reaction on and above the surface of the Earth. It is believed that there were no seas four billion years ago and there is adequate evidence to suggest that there were oceans much less than two billion years later. A very large part of this huge volume of water fell in the form of the condensate which we call "rain" during this period.

At first, because of the high temperature of the Earth's semisolid surface, the atmosphere built up from gaseous volcanic emission was incredibly hot. Because of the high water content of the emission, vapor pressure built up until one hundred percent humidity was reached at some point high above the Earth's surface where temperature and vapor pressure balanced. Continued volcanic emission of water vapor then resulted in the precipitation of rain at that elevation. As rain fell toward the Earth into the very hot region above its surface, the rain evaporated again, picking up latent heat and thus cooling the primitive atmosphere in its vicinity. This cycle continued until the surface of the Earth had itself cooled sufficiently to allow puddles of near-boiling water to form in depressions on its surface. This cycle of cooling was a process that probably continued unabated for hundreds of millions of years. Today, we have another name for a modified form of this process. We call it "weather." But so vast and continuous was the precipitation by condensation in primordial times that it quite likely led to an ice age early in the Earth's history.

Imagine the scene while all this was happening. Near-boiling streams, pools, lakes, and then cooler seas and oceans. One hundred percent humidity at a temperature around boiling point; thick, heavy, impenetrable clouds; tumultuous thunderclaps accompanied by instantaneous lightning; volcanoes belching glowing magma and steam. And centuries of torrential rain; rain that had absorbed some of the carbon and hydrogen compounds in the atmosphere to form acids.

This acid-rain reacted with the igneous rocks, extracting some minerals encrusted in the rocks. The sediments were swept down to the deepening seas by the torrents to form the first sedimentary rocks. By the very nature of their composition such rocks provide a record of their environment at the time of their formation. Collectively, sedimentary rocks are the Earth's filing system from which the chronology of its geological and natural history can be retrieved.

The oldest sedimentary rocks known on Earth were formed about three billion five hundred million years ago. These Precambrian rocks include the Banded Iron Formation, and major deposits of this kind of iron ore are recovered in the United States in Mesabi, Minnesota, and Cornwall, Pennsylvania. Banded Iron Formation is a beautiful rock peculiar to Precambrian terraces. It clearly reflects the primordial conditions in which it was formed. Strata were evenly laid down in open bodies of water and contained high concentrations of unoxidized iron, which suggests the mineral was leached out of igneous rock by surface drainage in a warm, humid, oxygen-free environment. Before free oxygen did become a significant proportion of the atmosphere, the pure mineral in the surface rocks was reduced to ferrous oxide. By dating the most recent time that unreduced ferrous irons appear in sedimentary rocks, the approximate time of change in the composition from an oxygen-free atmosphere can be deduced.

The Banded Iron Formation contains a sedimentary rock called "arkose." This is sandstone, a product of the erosion of granite. By definition, arkoses contain twenty-five percent or more of the mineral feldspar. In humid or wet conditions the feldspar content usually degenerates into aluminous clay, so the presence of arkoses in the Banded Iron Formation is interpreted by some geologists to mean that either desert or glacial conditions existed in Precambrian times.

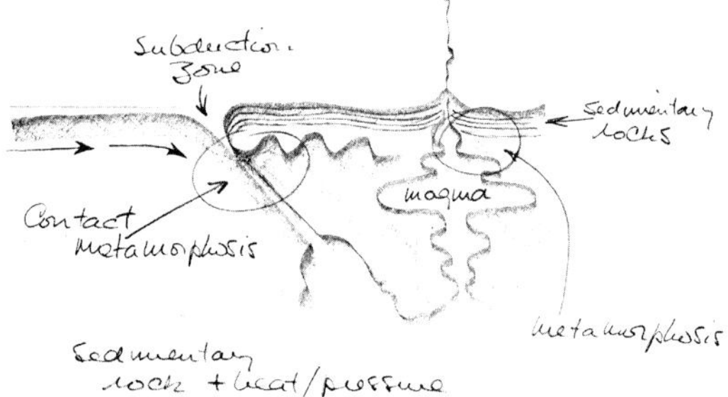

The convection currents in the Earth's mantle had their greatest effect on the crust while the latter was forming. The global wanderings of the first granitic continental masses might have been at their most vigorous at this time. When the first sedimentary rock had been formed from the accumulation of sediment on the first seabeds, some of it was eventually carried by sea-floor spreading to a subduction zone, where it was reprocessed. It was altered in nature not only by the subsurface temperature, but by compression. This resulted in a complete change in the form and composition of the minerals involved, a metamorphism. Such rock forms are called "metamorphic," a term used to describe any rock that has been fundamentally changed by heat and pressure and is subsequently re-exposed to the surface of the Earth by tectonic forces. Part of the subducted material was drawn back down into the asthenosphere with the seabed basalt to be reprocessed in the Earth's mantle.

There are considerable differences in appearance among igneous, metamorphic, and sedimentary rocks. Most igneous rocks consist of interlocking crystals, and their size is determined by the rate of cooling; the slower the rate the larger the crystal, and some, like obsidian, have no crystals at all: they are glass. Such rocks are generally extremely hard. Basalts and granites are the most familiar forms. When metamorphic rock forms, its crystals become aligned to accomodate the pressure and are often plate-like in shape. Such rock splits more readily than its igneous counterpart. Marble and slate are two of a great number of common forms of metamorphic rock. The character of sedimentary rock is determined by the size and composition of its particles. It can be composed of particles or fragments of pre-existing rocks, or accumulated organic matter, or crystals formed by solution or by chemical reaction. The three most common sedimentary rocks in the trinity of canyons are sandstone, limestone, and mudstone (see illustration pages 52–53).

On a worldwide scale, seventy-five percent of all visible rock is sedimentary. With the exception of Granite Gorge at the bottom of Grand Canyon, which was formed from igneous and metamorphic rocks, and the remnant volcanic fields in the general area of the Colorado Plateau which are entirely igneous, most of the rock to be seen in Grand Canyon, Zion, and Bryce Canyon is sedimentary in origin. One exception is a formation called "bentonite" which is a clay mineral formed by the chemical modification of volcanic ash and is found extensively in the Chinle Formation in the Painted Desert and at Zion.

Just as environment determined the character of the plateau canyons, environment also determined the exact nature of each layer of sedimentary rock found there, and for this reason it is possible to read that "character." The most essential elements of environment are an atmosphere, fluctuations in temperature, water precipitation, and angular slopes that enable the precipitation to act upon or react with surface rock. Given these basic conditions, four types of sediment accumulate: those on land, on the seashore, offshore, and deep in the sea.

Land deposits form alluvial fans at the foot of steep mountain slopes at a point where a stream of water loses its velocity. If the slope is at a sufficient angle, the sediment might be carried downstream while particles of sediment drop to the stream bed in proportion to the speed and volume of the water's flow. The nature of these deposits will depend on the origin of the water flow—which could be snowmelt, glaciermelt, or just heavy rainstorm—and the type of rock eroded.

A stream may run into a surface depression, in which case a lake will form, allowing some, if not all, of the sediment to precipitate. In arid conditions the lake will dry out and the sediment at the bottom will be exposed. Each succeeding rainstorm starts the process again, thus building up layers on the lake bottom. But in humid climates the lake may not dry out; it may fill the depression until it overflows. The deposits on the bottom will now accumulate steadily. The spillover of water will continue downslope until it joins a tributary of a river. Near sea level the speed of the river will slow, thus allowing further deposition of carry-over sediment at the river mouth. If the waterflow is insufficient to reach the sea, a marshland will form.

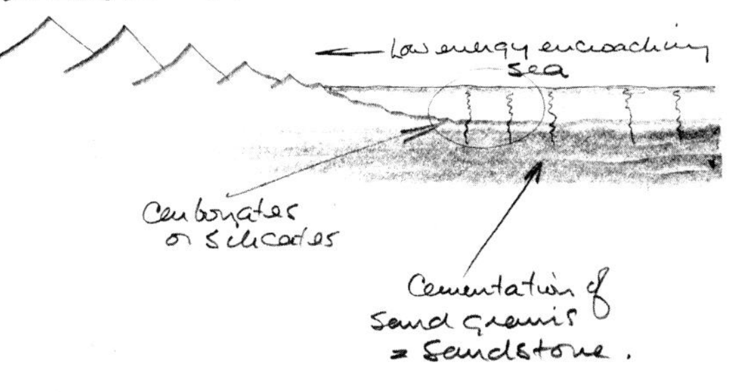

Suppose there is now an extreme change in global climate so that the whole area becomes hot, dry, and windy. The streams, rivers, swamp, and delta will gradually dry out. Particles of material not yet consolidated into sedimentary rock will be blown by the wind and become airborne. Wind-blown, or more exactly "eolian," sand dunes will form and shift with the prevailing wind. Desert conditions now prevail over the landscape. Let us say a few thousand years have passed and that a worldwide climatic change has melted the polar ice caps. The level of the sea will rise to cover the desert landscape to a depth of several hundred feet. We now have shallow marine conditions where once there were undulating sand dunes. The grains of sand became lightly cemented together by the presence of iron oxides, carbonates, or silicates washed into the sea. Other sediments will eventually form on the top of the now cemented and sea-covered dunes, and when these additional sediments accumulate to sufficient depth, the ancient eolian dunes will have become sandstone. Now let's assume that a further

Cross-bedded sandstone, Zion

Vast wind-blown ("eolian") sand dunes of Early Jurassic times (about 180 million years ago) have been etched by erosion—wind, summer storms, and winter snows. At lower levels in Zion Canyon the Navajo Sandstone, as the formation is called, forms natural arches and domes. The strange shapes to the left are "hoodoos." Harder rock surmounts the soft. The latter erodes, leaving a capstone precariously balanced on top.
11. D-3.

few million years have passed and that there is another change in the Earth's climate. The shallow sea remains until the ice caps form again. Meanwhile, sediment settles on the sea bottom over the newly formed sandstone.

As the polar caps form again the accumulating ice causes the level of the sea to fall. The seawater covering the new sandstone formation becomes shallow enough to allow evaporation to the point of producing a layer of calcium carbonate, and this builds up on the seabed to form yet another sediment which will become a stratum of limestone.

The cycle of sedimentation now repeats itself over millions of years. Strata accumulate one upon the other to a total thickness of several miles as the region gradually sinks, the uppermost strata never being far below sea level. As depositions increase in density, thickness, and total weight, the "waterbed reaction"—isostasy—plays its part by depressing the crust and maintaining our sedimentary formations under the surface of the sea.

Tectonic movement imposes variations on this theme and eventually these sedimentary rocks become subject to erosional forces. When a particular continental area like the one described is uplifted, faulted, and eroded, the result can be a spectacle of canyons like Grand Canyon, Zion, and Bryce, but it is environment that dictates the outcome. In the case of our trinity of canyons, an arid climate has predominated much of the time during which their rocks were eroded. In recent times, because of the high

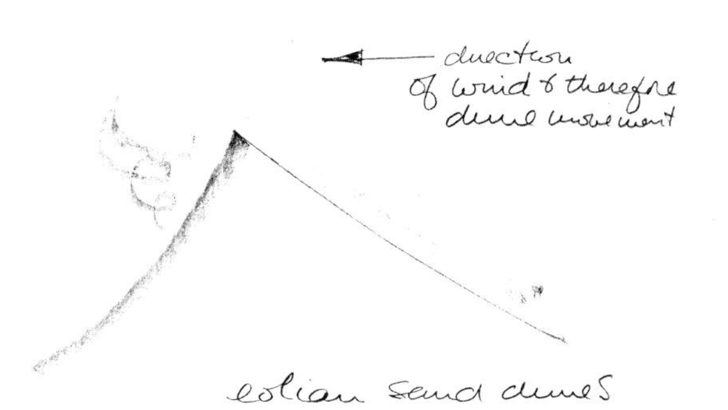

elevation of the Colorado Plateau, the area has been subject to violent summer thunderstorms and deep winter snows, but with very hot dry conditions in between. The formidable mountain range (the High Sierras) far to the west of the Plateau is the prime cause of its arid climate. Death Valley, the desert state of Nevada, and the Colorado Plateau are all in a "rain shadow," starved of Pacific moisture by this barrier.

Geologists are able to build a picture of ancient environments and climates from what they call "structures." The Navajo Sandstone structure at Zion is a near-perfect example. To qualify as a sandstone the particles that are cemented together must be of a specific size—between .0024 and .0078 in. If the particles are larger the rock is classified as a conglomerate. If the rock particles are less than .0024 in. the rock is classified as siltstone or mudstone, and if less than .0002 in. it is called a claystone.

The Navajo Sandstone exposed at Zion is an ancient desert of eolian sand dunes. That the dunes were formed by the wind from loose sand is shown by the manner in which their strata are formed with large cross-bedded sweeping surfaces. If the sand had been deposited in a stream, the cross section would show layers parallel to the surface of the water from which they are deposited. (Most stream deposits are also cross-bedded but are of different types.) A dune moves in the same general direction as the prevailing wind by the action of wind blowing up the slope to windward and over the crest at the top of the dune. The windblown particles then accumulate near the top of the steep downslope. The angle of the downslope is also the angle of repose of the sand, beyond which it slips or avalanches. By this action, the whole dune creeps in the same direction as the prevailing wind, and the sand on the downslope forms layers as it avalanches. Where the wind changes direction, new sets of sloping strata are formed on top of the older surfaces. The result is the characteristic eolian cross-bedded sandstone.

And at Zion such cross-bedding can be seen on a large scale (see panorama pages 58–59). A desert covered this area one hundred and fifty-five million years ago during the Mesozoic Era. Its sands were several thousand feet thick and covered thousands of square miles. It was as extensive as today's Empty Quarter in Saudi Arabia. Erosion has exposed the Navajo Sandstone at Zion and just as long as a surface is exposed, erosion will continue. Fine particles of sand loosened from the surface of the Navajo are blown off by the wind and sometimes carried for considerable distances. Some of the debris from the Navajo Sandstone cliffs in the vicinity of Zion (near Kanab, Utah) is carried and deposited by wind turbulence into an area called Coral Pink Sand Dunes (see panorama pages 60–61). These dunes are a present-day eolian desert, a reproduction in miniature of an ancient Mesozoic landscape.

There are many other kinds of sedimentary structures with characteristic bedding, ripple marks, mud cracks, raindrop marks, stream channels, deltas, sand bars, and so on. Each structure reveals its own story. It indicates not only the environment at the time of deposition, but also by comparison with similar modern environments, an indication of the time taken to form the deposit and the probable climatic conditions that prevailed when the sediment was laid down. Stratification is itself a sedimentary structure, and it can be assumed that the younger strata have been deposited on top of the older. But exactly how old is "old"?

There are two ways of dating any event. One can say "I was born during World War II" or "I was born on January 1, 1942." The first date is expressed in relative terms and the second in absolute terms. For much of its history, geological dating was almost entirely relative and this is why we have stratigraphic tables of eras, periods, or epochs, which were originally based on stages in the development of life. The advent of radiometric methods of determining age is gradually changing geological dating from relative terms into absolute terms.

The Earth's record of life is largely in fossilized form. As previously suggested, sedimentary rocks provide an orderly filing system. The filing system not only records environmental change, but also biological development, because strata are superimposed one on top of another in order of deposition. But whereas fossils provide an indisputable method of relating one formation of sedimentary rock to another, they do so in terms which are relative and not absolute. Fossils are correlators, not clocks. No fossil record can appear in igneous rock, and any previous fossil record is usually destroyed by heat during the process of forming metamorphic rock. There are comparatively few fossil forms in early Precambrian sedimentary rocks because, as one would expect, traces of life become vague as one goes back in time. And anyway, a large proportion of Precambrian sedimentary rock has been metamorphosed in the three thousand million years and more since life first appeared on Earth, so that the fossil record for this period has been reduced.

Before the discovery and development of radiometry, no one could say for certain how old the oldest recognizable fossils are, or how old the oldest rocks on Earth are, or even how old is the Earth itself. So the title "Precambrian" was coined to cover a time of unknown duration, the time before life—or so it was thought at the time the name was first used. The Earth proved to be far older than anyone could reasonably have guessed and now that this is known, the Precambrian Era is disproportionately long in comparison to its companion eras on stratigraphic tables.

The basis of radiometry is the unstable isotope. There are one hundred and three known chemical elements which can combine in an astronomical number of permutations to form every substance on Earth. Some of these elements are stable, others are naturally unstable. The unstable elements are called "radioactive." Many of the stable elements also have unstable twins; they and their twins are called "isotopes."

If left to their own devices, unstable humans tend to get more unstable, but chemical instability is a steadily "improving" condition. Either atomic particles are lost irretrievably by the parent unstable element, or the unstable element changes its character, a kind of alchemy. In the former case, the radioactive substance is said to have a half-life; it loses half its radioactivity in a given period of time. In the latter case, the parent isotope, as such substances are called, produces a daughter isotope, and the reproduction of a given quantity of daughter isotopes also takes a definite period of time. Therefore, the radioactive parts of both forms of unstable element decrease at a measurable rate in time.

The half-life of a radioactive substance can be determined very accurately. The time taken for it to decay to a particular point in its history can be measured in terms of years. Similarly, the rate at which a parent isotope produces daughter isotopes can be measured accurately, so the quantitative relationship between parent and daughter also indicates the lapse of time since reproduction began.

Some stable isotopes are primordial and are as old as the Earth. Others have

been formed at various intervals in the Earth's history as the by-products of radioactive decay. All unstable isotopes have varying but individually fixed half-life or reproductive capacities. Some have half-lives of billions of years, others hundreds of millions, still others millions; and the half-lives of some are mere minutes.

The moment these atomic clocks begin to tick is called the "time-zero event." In the case of igneous and metamorphic rocks the time-zero event is the moment of crystallization. In crystalline sedimentary forms like gypsum they begin when the crystals form an interlocked system, and in this respect are similar to igneous crystalline rocks. But if the sediments of which they are composed were organic and inorganic debris, then the time-zero event could be the formation of the rocks from which their debris particles were derived.

One would expect that radiometry could provide most of the answers to the problems of dating rocks, and of course in recent years it has established many dates which have caused a surprised eyebrow or two to be lifted in the world of geology. A recent estimate of the age of the oldest rock in Grand Canyon was made by using the potassium-argon method of radiometry, the isotope of argon being the offspring of its parent isotope of potassium. The samples of rock used for the estimate were obtained from the Granite Gorge of Grand Canyon, and analysis produced a different date from each sample. A variation of fifty to a hundred million years sounds considerable, but because it represents about five percent of the mean result, in fact the finding is pretty reliable. And how old was this particular bit of planet Earth? Just 1,700,000,000 years!

Coral Pink Sand Dunes

The Navajo Sandstone formation was formed from vast dunes. The Coral Pink Sand Dunes are a comparatively tiny area of dunes formed from windblown sand from nearby cliffs of Navajo Sandstone. They give an idea of what the original Jurassic dunes looked like.
12. D-4.

Queen's Garden

The spires of Bryce number tens of thousands, but perhaps some of the most impressive shapes are seen here in the Queen's Garden on the Navajo Loop Trail. The rocks at Bryce are among the most recently formed of the canyon trinity. They were formed of lime and sand accumulated here in the Tertiary Period of the present Cenozoic Era, about fifty million years ago. These pinnacles may be compared with those shown in the panorama of the Paria Amphitheater on pages 46–47. These are simply a more advanced form of the same erosion process.

13. C-4.

IV

THE LOST MILLENNIA

Storm in Grand Canyon

Taken in July above the Redwall Formation near Indian Gardens, Grand Canyon, this electric storm conveys a sense of the regular forces of erosion in the region. Static electricity made the author's equipment crackle and spark, and his hair stand on end while the picture was being taken.
14. F-4.

The mile-high cliffs of Grand Canyon reveal far more than one can see. If we imagine the formations which now make up the soaring crags of Zion and the pink confection of Bryce extending to the vicinity of Grand Canyon, another mile and a half of rock would be added to the Canyon's rims. Yet that would be only a modest reconstruction. If all the rock formations that are not represented in the Canyon, but which nevertheless were laid down in that region, were to be restored, the total would rise a further three or even four miles. Indeed, neither the rims nor the Canyon itself would exist. The fact is that much more rock has been eroded and washed away than remains to be seen in Grand Canyon today.

Most sedimentary rock is formed by the accumulation of deposits on a seabed. Whenever a sea in the Canyon region receded, the seabed formation was exposed to erosion. If enough time passed the formation beneath was also eroded to sea level. When another sea submerged the region new sediments were deposited on top of the eroded formations. There was then a considerable difference in age between the surface of the old rock and the underside of the new. Such incongruities between interfaces of rock formations are known as "unconformities." Unconformities reflect differences in the circumstances in which contacting rock surfaces were produced. Many of the unconformities in Grand Canyon, Zion, and Bryce are between horizontal formations and can be differentiated primarily because the fossils they contain are different. Sometimes when the formations both above and below the break are horizontal, an unconformity is evident only because some erosional feature in the lower

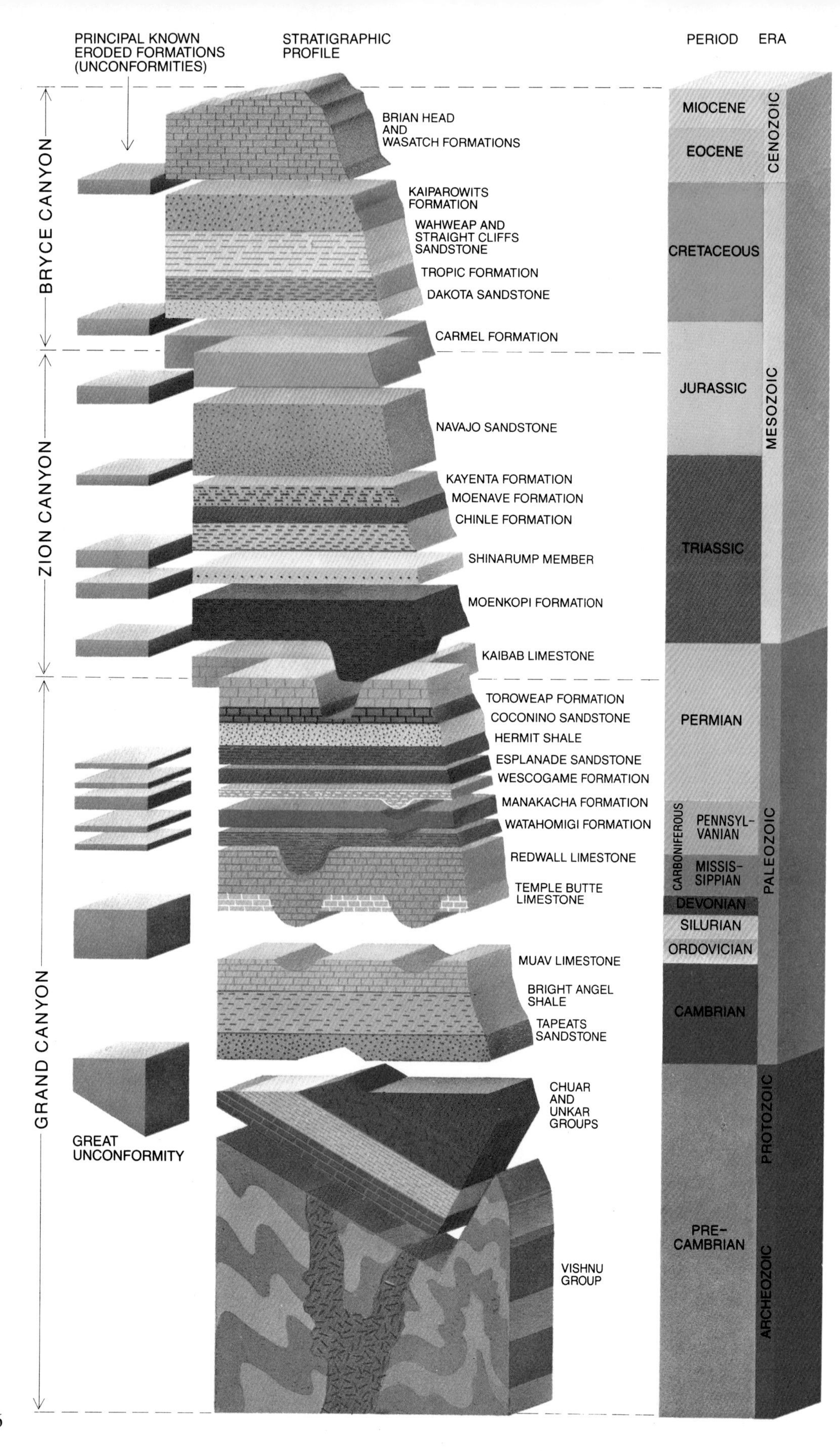

The Unconformities

When rock formations are eroded to near sea level and other sediments are later deposited which themselves become rock, there is an "unconformity"—a time gap—between the old and new formations. In some cases the unconformity can mean a period of hundreds of millions of years for which the rock record is missing. The Great Unconformity of Grand Canyon is an extreme case of this kind (see diagram at left). There are many other unconformities, some of them very substantial, between some of the rock formations in Grand Canyon through to the uppermost rocks of Bryce. Some of the major gaps in the geological record are shown here. If they had been drawn to the same scale as the other rock formations shown, their total thickness would be greater than the overall diagram.

Unkar Creek, Grand Canyon

The Colorado River here sweeps through 90 degrees from Marble Canyon into Grand Canyon. This panorama is taken from the west and looks toward the Palisades of the Desert at its center. The low flat area at right was inhabited centuries ago by Anasazi Indians. Most of the rock formations that can be seen are Precambrian, and are among the oldest sedimentary formations in the Canyon, perhaps part of the continents which preceded the formation of the supercontinent of Pangea. The rocks that form the horizon to the center and right are comparatively young. The unconformity between the two can clearly be recognized by the meeting of the angled and horizontal formations. Unkar Rapid, center right, rates 7 on a scale of 1 to 10 of difficulty for boatmen.
15. F-4.

formation, such as an old river channel, is filled with sediment of later deposits. On the other hand, erosion may have planed down the original surface to a flat surface—called a "peneplane"—before new deposition commenced, or the new sequence of stratified sedimentary rocks may lie on top of igneous or metamorphic rocks that may have been leveled. When the underlying sedimentary rocks have been tilted before peneplaning and redeposition, the relationship is called an "angular unconformity." These last characteristics, peneplanation and angular unconformity, are the dominant features of the Great Unconformity of Grand Canyon. The Great Unconformity is among the most fascinating of all natural phenomena to be seen on the Earth's surface (see panorama pages 15–18).

Almost anywhere in Grand Canyon west of Bright Angel Canyon, two contiguous rock formations, which have a most considerable gap in time between their interfaces, can be seen at the top of Granite Gorge (see panorama pages 8–9). The lower of these two rock units is the Vishnu Group, which consists of a mixture of granites and schists. The Vishnu forms the steeply inclined walls of Granite Gorge rising in some places about a thousand feet from Colorado River level and its rocks are about one thousand seven hundred million years old—the oldest rocks in the Canyon. Their uppermost surface was worn to a peneplane by erosion before horizontal formations of Tapeats Sandstone were deposited on it in early Cambrian times one thousand one hundred and thirty million years later. Millennia of millennia of lost millennia separated the two events. What could have happened during this incredibly long period of time? The clues that allow a reasonable deduction to be made are right there near the bottom of Grand Canyon.

About two thousand million years ago, the rocks that were later metamorphosed into the Vishnu Schists of Granite Gorge were sedimentary rocks formed from deposits of sand and mud and igneous rocks of granitic type. During subsequent tectonic upheaval they were bent, deformed, compressed, and folded as the continent heaved in response to the driving force of continental drift. Even as the mixture of igneous and sedimentary rocks was metamorphosed by this process it was penetrated by liquid granite magma from the underlying asthenosphere. The general result of this upheaval was a mountain range probably snowcapped and of considerable extent and of alpine proportions (see illustration pages 80–81).

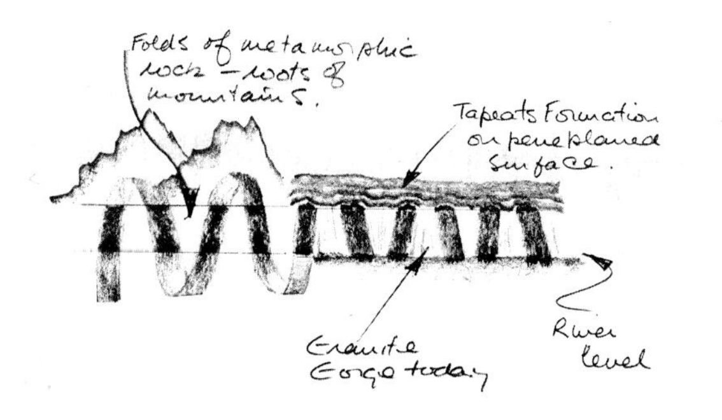

That there were sizable mountains in the region is evident from the character and the nature of the Vishnu Group of granites and schists in Granite Gorge. When one travels downriver by boat through the Upper, Middle, and Lower Gorge, one can see what is left of gigantic folds in the rock, folds that are on so large a scale that they appear to be no more than a series of unrelated, near-vertical columns. But they are related in the sense that split pairs of similarly composed columns of Vishnu Schist and granite radiate from a central core. This pattern repeats time and time again along the whole length of Granite Gorge. The columns are similar in character to the roots of mountain ranges that can be seen elsewhere in the world. They are composed of dense crystalline rocks of types commonly associated with the extreme pressures generated during mountain building (see illustration pages 76–77).

Mountain structures similar to those of the Precambrian alpine range that dominated the landscape of Grand Canyon about seventeen hundred million years ago can easily be visualized because there are many spectacular examples of similar mountain ranges in existence today; for example, the European Alps, the Urals, and the Carpathians. These are relatively young mountains and it is not easy to imagine them being reduced to peneplanes, although that is their destiny.

The shape of mountains depends in part on how they were formed. Mountains which are folded and elevated by the pressures created by continental movement, such as those presumably represented by Vishnu Schist roots in Grand Canyon, generate the very conditions that lead to their own destruction. The higher they are elevated the more fragmented the topmost folds become as internal stresses are relieved. And as they rise, such mountains create their own weather (see illustration pages 80–81). The greater the altitude they achieve, the greater the probability of snow precipitation on their summits. As snow accumulates it packs under pressure and compresses into ice which is plastic. Glaciers—rivers of ice—flow down mountain flanks, grinding rock surfaces and carving off large volumes of rock that are transported to the snout, the point at which glacier ice melts. Glaciers are among the most destructive forms of water erosion. In ten thousand years a glacier can carve a valley from extremely hard rock. Erosion by an equal volume of free-running water might take millions of years to cut a valley of similar proportions in the same rock.

Although no evidence of glacial action has been discovered in the Vishnu Schists and granites of Grand Canyon, evidence of substantial glacial activity has been found in the Headquarter Schists of Wyoming five hundred miles due north. From ancient glacier debris it has been proposed by G. M. Young that between sixteen and twenty-four hundred million years ago glaciers similar to the modern glaciation of Antarctica extended to that region and reached a depth of twenty-five hundred feet or more. Young believes that a vast ice sheet stretched from the Great Lakes to the mountains of Medicine Bow in northern Colorado.

Young's findings at least tentatively suggest that the Precambrian mountains of the Grand Canyon might have existed at a time of severely cold climate, in which case the likelihood of heavily snowcapped peaks with attendant glaciers is strengthened. Glaciation or not, the freeze–thaw cycle associated with high mountains in any climate, allied to other forms of mechanical erosion, probably played a dominant role in peneplanation. Chemical weathering—the dissolution of soluble materials, the reduction of minerals into clays by the action of water, carbonation, oxidation, hydrolysis, and so on—played its part too. Time did the rest—hundreds of millions of years of time, until the mountains were no more, and the remnant peneplane was reduced almost to sea level.

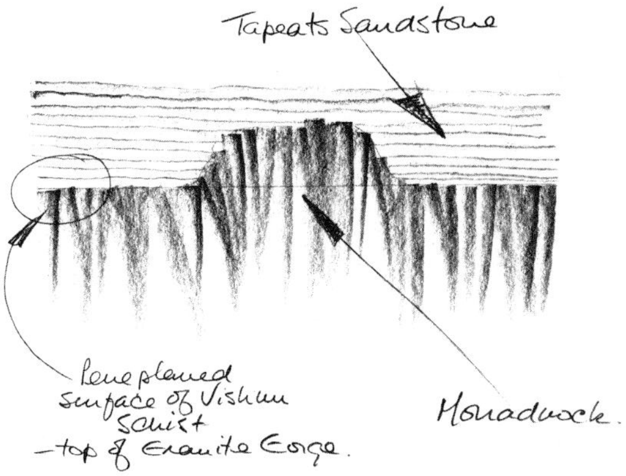

From the South Rim the peneplaned surface of the Vishnu Group at the top of the Granite Gorge looks even and flat where the Tapeats Sandstone covers it (see panorama pages 70–72). This is an illusion of distance. Not only does the Vishnu surface undulate, but it frequently protrudes into the overlying Tapeats which has formed sedimentary layers around the protrusions. These protrusions, often several hundred feet high and stretching over considerable distances, were once wave-battered islands in a sea that gradually invaded the Vishnu peneplane. They are called "monadnocks," a name derived from Mount Monadnock in southwestern New Hampshire which is a classic example of a conspicuous isolated hill of hard bedrock that has resisted the erosional forces that reduced the surrounding terrain.

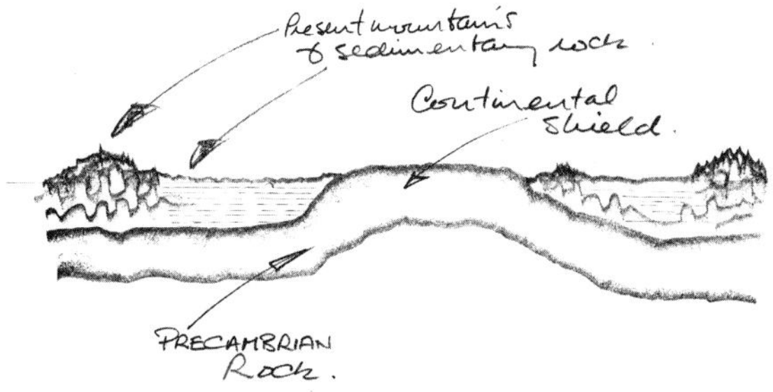

The story now moves to the eastern end of Grand Canyon to the region of Unkar Creek, which is so magnificently displayed for all to see below Lipan Point on the South Rim. I like to think of Unkar as the "engine room" of the canyon because of the panoply of momentous geological events that occurred in its vicinity (see panorama pages 15–18). It was perhaps just east of here that in comparatively recent times the Ancestral Colorado River was stolen by the Hualapai drainage system and its path diverted to its present course. This is where a section of the forming primordial continent of Pangea in Paleozoic times can best be viewed. But long before these events, the ancient continental shields—the original granite masses, which underlie the present continents, now encrusted with sedimentary rocks—had grouped and perhaps regrouped several times, as they accumulated mile upon mile of sedimentary rock on their surfaces. At Unkar Creek that remote time of geological history is partially revealed, for some of those ancient Precambrian sediments are here. The view is breathtaking. The geological events are mind-boggling. The combination is irresistible.

The downward-inclined terra-cotta-colored rocks stretching from west to east across the wide floor of Grand Canyon below Lipan Point, with Unkar Rapid at the center, consist of the six formations of sedimentary and igneous rocks that make up the Grand Canyon Series. These are the Precambrian rocks that once formed the landscape of the primordial continent that preceded Pangea. The sedimentary formations of the Paleozoic Era lie unconformable on top of these rocks. To me the scenario

After the storm, above Indian Gardens

Grand Canyon is a place of rapidly changing moods. This panorama was taken only an hour after the panorama on pages 64–65, at the opening of this chapter. Dramatic erosion in this arid region is the result of such violent and sudden storms. In temperate climates the changes are steady but persistent. This panorama gives the best view of the Tonto Platform, which is mentioned in later chapters. It dominates the lower and middle foreground here, and a mule trail can be seen crossing it at left. The most prominent features are the butte known as the Battleship at the extreme left, Cheops Pyramid at center left, Bright Angel Canyon at top center, Zoroaster Temple at center right, and O'Neill Butte and Yaki Point at far right.

16. F-4.

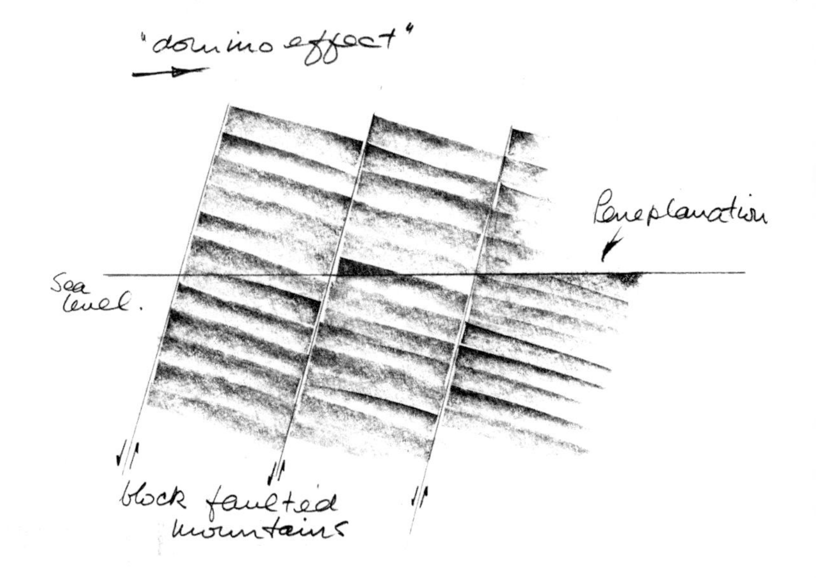
"domino effect"
Peneplanation
Sea level.
block faulted mountains

is almost inconceivable. At Lipan Point one is standing on what we accept to be the present North American continent yet below is a glimpse of the forming continent Pangea, which originated nearly a thousand miles away, and below that is a section through a third continental landmass.

Before transferring the scene to the Unkar region, we left the peneplaned Vishnu remnant of the Precambrian alpine mountain range on the point of subsiding below the sea. The subsidence continued with brief intervals of partial uplift and erosion until the Grand Canyon Series of sedimentary rocks had accumulated on its surface to a thickness estimated to be about fifteen thousand feet. When the process of sedimentation came to an end, a prolonged period of uplift commenced.

With this enormous mass of sedimentary rock to support, it was inevitable that when upward crustal movement occurred, tension would develop in the underlying crust. In this instance the strain was relieved by block-faulting. As a consequence, a range of block-faulted mountains was formed with Vishnu Schists and granite as its base. This type of mountain, often abruptly angular, sometimes achieves high altitude, and a "domino effect" causes each gigantic block of rock formations and basement rock to be tilted at an angle to its neighbor. In this case, the upper part of each block that protruded above sea level was eroded, and eventually reduced to a peneplane. The underlying parts which had been pushed downward by tilting into the Earth's crust were protected from erosion. Because of the angle of inclination of the blocks, the surface features of the peneplane were not homogenous. They were a mixture of Vishnu Group and Grand Canyon Series rocks. This second great Grand Canyon peneplanation took millions of years to complete before the terrain was once again at sea level and therefore susceptible to invasion by the sea. The first marine deposits which followed were of sand, the sand which became the Tapeats Sandstone Formation.

From either the South Rim or the North Rim in the vicinity of Bright Angel Canyon, one can look down on a considerable rock bench above the Tapeats Sandstone that is called the Tonto Platform, which for the most part directly overlays the Vishnu Group (see panorama pages 70–72). Looking east along the South Rim from Lipan Point (see panorama pages 36–37), one can look back along Granite Gorge toward Bright Angel Canyon. From this perspective only part of the Tonto Platform can be seen, but one can see the Tapeats Sandstone in section lying directly on top of the peneplaned Vishnu Group. A wedge—the Grand Canyon Series—begins to appear between the two from west to east and this grows steadily in extent until, at the head of Upper Granite Gorge at the eastern end of the Canyon, the whole spectacle is transformed into the angled formations which typify the Unkar Creek region of Grand Canyon (see panoramas pages 66–67 and 15–18).

Where the Tapeats rests directly on the surface of the Vishnu Schist, the unconformity represents a gap in time of twelve hundred million years. During this immense period of time, three major geological events took place. First, a complete range of alpine-type mountains was formed and eroded to a peneplane. This episode was followed by the deposition of the Grand Canyon Series with a one-time total thickness greater than that of all the present-day formations in Grand Canyon, Zion, and Bryce canyons combined. Finally, this series was uplifted, converted into a mountain range, and partly eroded away. In the light of these tremendous events, it makes one wonder what else of significance could have happened. A great deal.

Any stratigraphic chart of Grand Canyon has two significant gaps that are easily overlooked. They represent the Ordovician and Silurian periods of the Paleozoic Era. These are the geologic episodes named after the ancient Welsh tribes who were thorns in the sides of the Roman armies that invaded the British Isles many centuries ago. The Ordovician and Silurian periods cover one hundred and five million years of geologic history. In order of deposition, they followed the Cambrian Period (Muav) and preceded the Devonian (Temple Butte) Period. Scientists are faced with a choice when they recognize an unconformity of this character. They have to decide if the missing rock sequences have been removed by erosion, or if they ever were deposited in the region at all. Most Ordovician and Silurian rocks are known to have been formed primarily from deposits laid down in shallow seas. If the rocks that are missing from Grand Canyon were never formed within its province, it could only mean that the regional landscape was elevated above sea level throughout Ordovician and Silurian times so that no marine deposition was possible (see chart page 66).

During a period as long as a hundred million years—the approximate duration of the Ordovician and Silurian periods—erosional processes are bound to leave distinctive marks clearly incised into the final surface. When that surface eventually receives a deposition, the erosion channels may be filled with a sediment that can be dated. In effect, a new sedimentary rock has keyed into the incised portion of the older

Mountain-building

Folding-and-thrusting ("orogeny") is one of several ways that mountains form. The mountains which once might have had the ancient schists of Grand Canyon as their roots were probably formed in this way. (A panoramic reconstruction may be seen on pages 80–81.)

First, sediments, silt, mud, and so on were formed under the surface of the sea. Either seabed subsidence or rising sea level allowed strata to develop layer upon layer until compression or chemical change caused rock to form.

Eventually, these sedimentary formations were buried deep in the Earth's crust, which caused their metamorphosis into schist by heat and pressure. Tectonic movement caused the schist to fold, uplift, and fault.

formation. In Grand Canyon such interfaces of erosion channels are found between rocks of the Cambrian and Devonian periods and between rocks of the Cambrian and the Mississippian periods which followed the Devonian. This fact has been interpreted by many geologists to be strong evidence that throughout Ordovician, Silurian, and much of Devonian time, Cambrian rock surfaces were continuously above the level of the sea. There is an opinion counter to this reasoning, which is simply that the missing rocks were indeed formed, but were subsequently so completely eroded as to leave no trace at all. In fact, a recent seismographic section of a considerable area of the North Rim of Grand Canyon is thought by proponents of this latter theory to indicate some presence of the missing formations.

While the rocks of the Ordovician and Silurian periods are entirely missing and there are only pockets or channels filled with lavender-colored Devonian Limestone, (called Temple Butte) in the eastern half of Grand Canyon, there are high cliffs of Devonian rocks in the western half. The great Mississippian Redwall Limestone is the most prominent vertical cliff to be seen as a continuous formation from one end of Grand Canyon to the other, and is the next formation up from the Devonian Temple Butte. Where the Temple Butte is missing, the Redwall lies directly on the Cambrian Muav Limestone. The time of unconformity or hiatus between the foot of the Redwall and the underlying rocks, therefore, ranges from about twenty-five million years to about one hundred and fifty-five million, depending on which surface is contacting which for the whole length of the Canyon.

Simultaneous with the formation of mountains, erosive forces began to act on the landscape. Magma was injected into the faults, producing granite intrusions, basalt dikes, and surface lavas through volcanic vents.

Finally, erosion reduced the mountains to a peneplane near sea level. A few "monadnocks" remained—rocky hills that had resisted erosion. Sedimentation began once more as the land was submerged by sea.

The height of the Redwall (about six hundred feet) represents only a part of the Mississippian Period which spanned twenty-five million years in total. Because much of the Middle Pennsylvanian above it (represented by the lower part of the Supai Group) is also missing, there appears to be an unconformity between the Redwall and the overlying Supai amounting to sixty million years. Erosional valleys have been cut into the upper section of the Redwall, and it is the present-day rock content of these valleys that provides evidence of a substantial historical break. This break is one of the most recent and exciting discoveries in Grand Canyon. Its discovery and explanation is a classic example of geological interpretation.

George H. Billingsley, a geologist of the Museum of Northern Arizona in Flagstaff, has explored many previously unvisited canyons in the western end of Grand Canyon. In addition to a well-developed sense of adventure, for some of the places he has visited would make intrepid bighorn sheep hesitate, he has been preparing data for a geological map of the western section of the Canyon. In September 1978 he gained access to a side canyon in the Diamond Creek area where he noticed what he thought might be a stream-cut valley in part of the Redwall Formation. It appeared to Billingsley that the valley was hundreds of feet deep and had been filled with unknown river sediments, capped with the marine limestone underlying the Supai Group.

Some months after his discovery, Billingsley returned to the scene by helicopter with an expert interpreter, Dr. Edwin McKee. Measurements and observations were made. Sample rocks and fossils were taken and over the months that followed it was determined that the Redwall Limestone—a marine deposit—had been raised a total of four hundred feet above sea level by crustal movement soon after the sea withdrew in the middle of the Mississippian Period. During this time and in the locality of Billingsley's find, a freshwater stream had carved through the thickness of Redwall on its way down to the sea. As the Redwall continued to uplift, the cove was enlarged by the combined action of the river and sea until the channel reached a depth of four hundred feet. At this point uplift slowed and stopped. Freshwater stream deposits collected at the bottom of the inclined streambed. The deposits included fossilized ferns and other plants of the early Pennsylvanian Period. Eventually, the Redwall Formation sank below sea level and a period of marine deposit began.

The effect of a discovery of this kind—a previously unrecognized crustal uplift, the erosion of a deep valley, followed by sea transgression—is locally profound because it dictates reappraisal of the record of the rock formations involved. Knowing for sure that a major event occurred at a particular time makes a radical difference in the way in which geologists seek further information relative to that event. But as exciting as Billingsley's discovery may be, in the context of the geologic history of Grand Canyon, it is just one more piece put into place in a jigsaw puzzle which is far from complete.

There are many more major unconformities between the strata of Grand Canyon, of Zion, and of Bryce Canyon. These are shown, as far as scale will allow, in the stratigraphic chart that appears on page 66. All were caused by the erosion of surface areas above sea level and were followed by more deposition under the sea, or on the bed of a lake or river, or by shifting desert sands. The most noteworthy of many, all of which were millions of years in duration, include that between the Kaibab

Limestone, which forms the North and South Rims of Grand Canyon, and the first strata of the overlying Moenkopi Formation which is the basal rock of Zion, and another between the top of the Moenkopi and the overlying Chinle Formation.

There is yet another big gap in the record—much of the Upper Jurassic and the Lower Cretaceous rocks are missing—between the Entrada Sandstone that forms the top strata visible at Zion, and the Dakota Sandstone basal rocks of Bryce Canyon. And finally, at Bryce Canyon most of the Paleocene Epoch rocks of the present Cenozoic Era are missing. They should lie between the Kaiparowits Formation and the Wasatch, which forms the pink cliffs, eroded spires, and monuments that typify Bryce. That represents a ten-million-year gap in the geologic record, a hiatus that occurred about fifty-five million years ago.

Above the surface of the sea, erosion in one form or another has not stopped for an instant since the day in primordial times when that first raindrop fell; it has only varied in degree. A gravestone can lose its inscription in a few decades, a stone building its pristine grandeur in a few centuries, and the Sphinx its façade in a few millennia. Mountain ranges take a little longer, and low-lying continents longer still. The initial height and angle of inclination of the surface feature above sea level determines its rate of decline.

Most of the southern Colorado Plateau is at an elevation between six and ten thousand feet. During the last ten million years large parts of it have been removed. Before another ten million years elapse, most of it will be gone. No Bryce, no Zion, no Grand Canyon; just a low-lying peneplane relieved by an occasional monadnock. All will be ready to disappear below a high-energy sea—the Pacific Ocean—and to receive the next deposit of sediments. It is perhaps difficult to perceive, but the present is a time of rapid erosion, a time of future unconfomity. The great canyons of the Plateau Province are here today but will certainly be gone tomorrow.

Alpine mountains rooted in Granite Gorge

Some geologists believe that the metamorphic schists and igneous rocks of Grand Canyon's Granite Gorge are the roots of an extensive alpine range of mountains. If this is so, they existed about 1,700 million years ago in far off Precambrian times. This is known by the radiometric dating of the schists. The panorama was taken from Havasu Point (South Rim) on a cloudless day so that this artist's reconstruction could be produced. From the size and character of the near vertical formations in Granite Gorge it has been estimated that the mountains were between fifteen and twenty thousand feet in elevation. While such mountains would have been snowcapped at least during the winter, it is impossible to establish whether or not they were glaciated. 17. E-4.

V

OF TRILOBITES AND BRACHIOPODS AND GREAT RED WALLS

View from Shoshone Point

Each of the many layers of sedimentary rock formations in this panorama can be dated by the "index fossils" they contain, of which certain species of trilobites are good examples. At the left center can be seen a 25-mile stretch of the Kaibab Plateau (North Rim), curved as a result of the uplift that contributed to the formation of Grand Canyon. Cheops Pyramid is at far left, Zoroaster Temple at center left, with Lyell Butte in left foreground. At center right can be seen the peak of Vishnu Temple, with the Palisades of the Desert, Desert View, and Lipan Point to the right. 18. F-4.

In sixteenth-century Europe the strength of a new opinion was often put to a simple test. The iconoclast was offered a choice between renunciation and incineration. But sometimes the heresy was considered so profound that no option was given. The inquisitors denounced, the stake beckoned, and the victim was burned alive. Such was the fate of a French Huguenot named Bernard Palissy at the Bastille in 1589, and an Italian astronomer-mathematician, Giordano Bruno, in Rome on February 17, 1600. Palissy was a potter, and was familiar with quarries and clay pits and therefore with the existence of fossils. He gave lectures on natural history that were published in book form in Paris in 1580. His main theme was that such fossils were the remains of living creatures. This view hastened his untimely end. Bruno was far less circumspect; he was not only a vociferous advocate of the Copernican theory, but openly questioned some contemporary Christian beliefs. Not content with questioning the occurrence of the Great Flood, he suggested that the sea had often invaded the land in the past. He published his ideas, also in Paris, in 1585.

At the turn of the sixteenth century the inflexible belief of most Europeans was that the catastrophic Deluge alone was responsible for all changes on the face of the Earth since Genesis. The views of Palissy, Bruno, and their kind were intolerable to the vast majority. Palissy and Bruno led the way to the scientific rediscovery of the basic principles of paleontology and stratigraphy first developed philosophically by the Greek scholars of the sixth century B.C., Xenophanes, Xanthus, and Herodotus. It was these scholars who first proposed that all present Earth processes are similar to those that occurred in the past, a philososphy known as "uniformitarianism." It was they who suggested that the presence on land of marine fossils implied that seas had once been greater in extent. But a thousand years passed before these revolutionary ideas took root, and then they required the perception of a man of unparalleled genius: Leonardo da Vinci, who rediscovered the forgotten concept of uniformitarianism. He also understood both the broad implications of the remains of once-living marine faunas and sedimentary rocks, and the relationship between them.

The science of geology was conceived by Greek scholars and born of Leonardo da Vinci's perception. It was brought into practical effect exactly one hundred and fifty years after da Vinci's death with the publication in 1669 of "The Prodromus of Nicolaus Steno's Dissertation Concerning a Solid Body Enclosed by a Process of Nature Within a Solid." In this treatise Steno not only laid down the foundations of the science of crystallography, but followed da Vinci's lead and pronounced on the relationship between fossils, sediments, and rocks.

Steno was born in Copenhagen in 1638 (his original name was Niels Stensen) and by 1665, while physician to Grand Duke Ferdinand II in Florence, he had developed a strong interest in the natural sciences. There is little doubt that Steno knew all about the geological martyrs Palissy and Bruno but, while sharing their views, he had no intention of sharing their fate. So, rejecting the idea of his time that mountains grew out of the ground like trees, he suggested instead that they were made of sedimentary layers formed in chronological order. However, he was careful to conform to the Church's timetable for these events by suggesting that the order of appearance of fossils might be related in some way to the history of the Earth since Genesis. He carefully avoided any indication of a period greater than the Biblical six thousand years and proclaimed that fossilized bones, which he had found during his travels, were obviously those of one of Hannibal's elephants.

Some of Steno's conclusions were so profound that they became the basic laws of stratigraphy. He established that because of their fluid state during formation, undisturbed horizontal beds of sedimentary rock always lie upon each other in chronological sequence and the lowest strata are therefore the oldest in a series. From this he reasoned that irrespective of the subsequent angle of underlying strata, a new stratum will always be horizontal when formed. And he concluded that unless it abutted against another rock face, a stratum would never have clean-cut, vertical edges at its extremities; it would just thin out and vanish. So in a series of sedimentary rocks the oldest strata will be at the bottom; if exposed strata are not horizontal but lie at differing angles to each other, there must be an unconformity between them; and if the edge of a series of strata can be seen, that edge has been produced by subsequent and extraneous events. It wasn't long before geologists began to document the thickness, sequence, and type of strata, and by the late eighteenth century they had developed a classification of rocks into eras and ages, using the terms "primary," "secondary," and "tertiary." This system was the forerunner of the present classifications of "era," "period," and "epoch." The use of words to classify periods of time had the distinct advantage of allowing geologists to think, discuss, and publish their findings without getting involved in either antediluvian controversy or commitment to precise chronology; they were not committed to specific dates, but only to an order of occurrence.

By the turn of the eighteenth century many investigators were probing the

Sea level

The oldest sedimentary rocks are at the bottom of a series. Only rarely can they form with vertical edges.

mystery of the Earth's evolution. Those who most influenced the advance of geology, particularly in biostratigraphy—the identification of strata by using fossil references—were Baron Georges Cuvier, Jean-Baptiste de Lamarck, and William "Strata" Smith. Cuvier laid the foundations of modern paleontology—the biology of past geological ages—and proclaimed that fossils demonstrated successive epochs in the Earth's history. Lamarck deduced that organisms had evolved as a consequence of environmental changes. And Smith recorded the consistent relationship he had observed between certain strata and a succession of fossils that were each slightly different from their predecessors.

From Steno to Smith, and then from James Hutton, Sir Charles Lyell, Charles Darwin, and a host of other nineteenth-century scientists, there followed a profusion of discoveries fundamental to our present understanding of the Earth. All were based on the philosophy of uniformitarianism propounded by Xenophanes, Xanthus, Herodotus, and advanced by da Vinci: The present is the key to the past.

Today we know with reasonable certainty that the Earth was formed about four billion six hundred million years ago; far, far older than any scientist contemplated even half a century ago. We also know that about four billion species of living things have evolved to date, many times more than the wildest guess of a few generations ago. The passage of time and the degree of evolution are organized and quantified in geological terms by determining the order of rock stratification and its associated fossil types. An unknown stratum can be dated by relating its fossil content to known evolutionary order, and an unknown fossil species can be placed in its correct evolutionary position by its occurrence in a known strata.

Fossils abound the world over and take many forms—sometimes rather unexpected forms. For example, smooth round pebbles which are found in the rib cages of excavated dinosaurs are called gastroliths. These stones possibly performed the same function for dinosaurs that gravel does for chickens today, and were simply gizzard stones which assisted primary digestion; or maybe dinosaurs were just clumsy in their eating habits. Whatever their purpose, gastroliths count as fossils. Excrement is another unlikely candidate for the fossil list; specimens are called coprolites and are highly informative digests of their now-extinct depositors' dietary habits.

Paleontologists sometimes find fossilized animals preserved in an almost complete state: sloths in arid caves, mammoths packed in ice, and men in peat bogs. Such effective preservation was the result of rapid reduction of moisture content or temperature, impregnation with chemicals, exclusion of air, or of a mixture of all four. Although we would hardly call preserved food "fossil food" when we buy it from a supermarket, there is really nothing new about desiccation, deep freezing, chemical additives, vacuum packaging, and various combinations of all four.

But by far the most common forms of preservation have been by carbonization and petrification. Carbonization is the result of incomplete decay in the absence of air succeeded by pressure. The organic content of the plant or animal is the only remnant, and takes the form of a tissue-thin trace of a fish or animal embedded in sedimentary rock or perhaps a massive seam of compressed carbonized wood and vegetable matter called "coal." Petrification took place when mineralized water wholly or partially dissolved the hard parts, such as bone, shell, coral skeleton, or wood, which were very

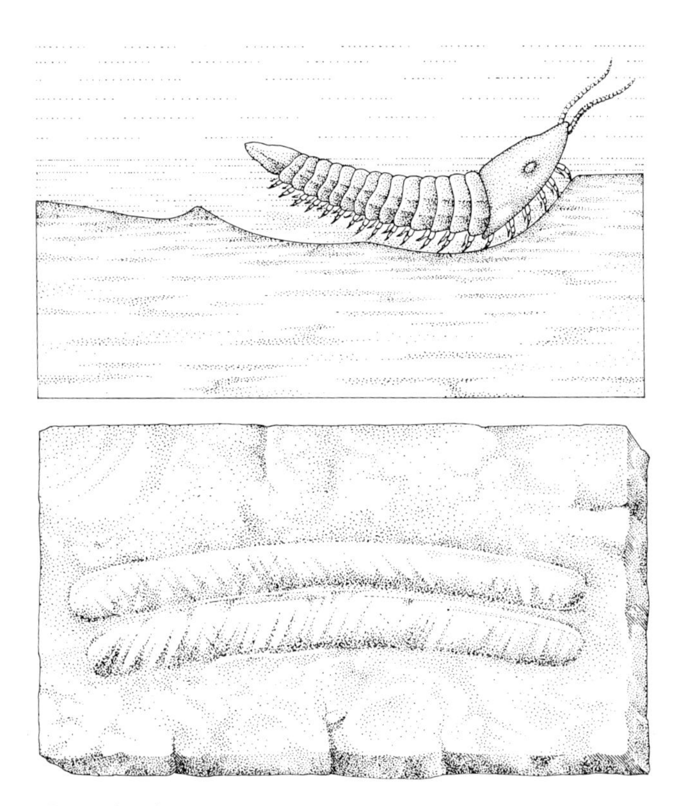

Trace fossils

The pairs of drawings on this page, and on the facing page above left, represent trilobites and the different fossilized tracks they have left. Immediately above, a trilobite is seen crawling, with the kind of tracks left behind as a result. Above right, a trilobite is shown foraging for food, along with its circuitous tracks. The trilobite shown on the facing page is burrowing for food, and the companion drawing shows the raised pattern of the burrow on the ocean floor. There were many species and subspecies of trilobites, and they varied in size from a fraction of an inch to 30 inches. They dominated the Cambrian seas for some 350 million years before becoming extinct.

slowly replaced by minerals in solution within the sediment in which they were buried. In some examples such replaced fossils perfectly duplicate even microscopic details of the original structure and form. It is rare for the original shell, bone, or wood to remain unchanged. But it is common for the internal structure and parts to have been destroyed, while the external form remains unchanged.

Sometimes the original plant or animal left only an impression of its presence: a footprint, a burrow, or a cast, and these are known as "trace fossils." In fact, some trace fossil animals are known to have existed only by the tunnel systems which they left for posterity. Various species of these creatures, the *annelida*, are named after the shape of the uniquely patterned warrens that they burrowed.

A perfect fossil that has retained details of the original form and structure of its skeleton may provide a paleontologist with an exciting opportunity to reconstruct intimate details of the donor's anatomy and physiology. Although encyclopedic, the fossil record provides little more than a glimpse of the volume and extent of past life. Then as now, most living things just died and decayed leaving no trace at all, or their remains were wholly or partially eaten by predators, or were weathered, worn, crushed, or ground up by pounding waves. Geologists, therefore, commonly must interpret the environment and age of sedimentary rocks from imperfect fossil fragments and their assemblages. Moreover, geologists generally have to segregate the clutter of extraneous fossils from those which are pertinent to the strata being investigated. The erosion of previously overlying and therefore younger strata may misleadingly have caused the redeposition of fossil remnants at a lower and older level.

During the course of geological research, hundreds of specimens will be identified and recorded from carefully measured locations within the boundaries of a formation. As data are correlated and the upper and lower horizons of occurrence of

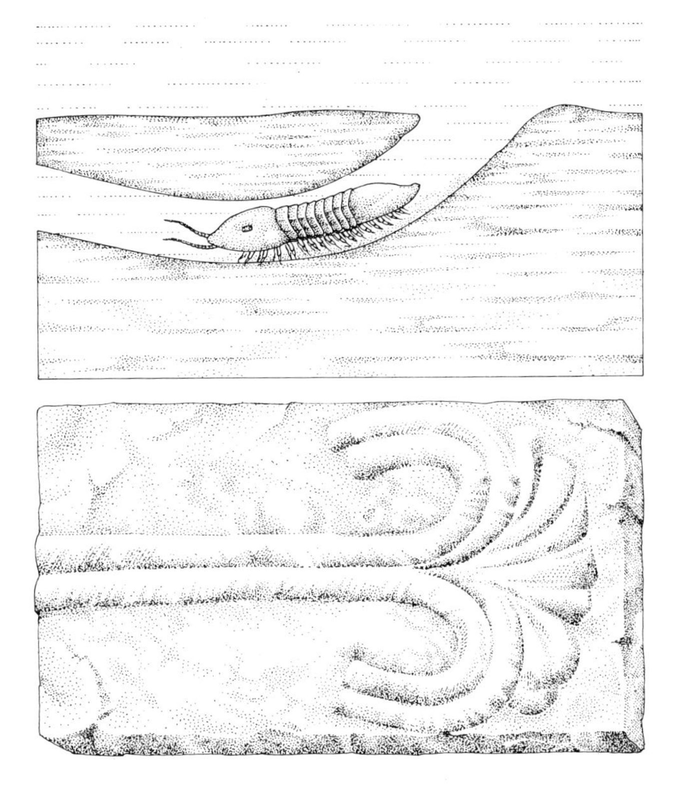

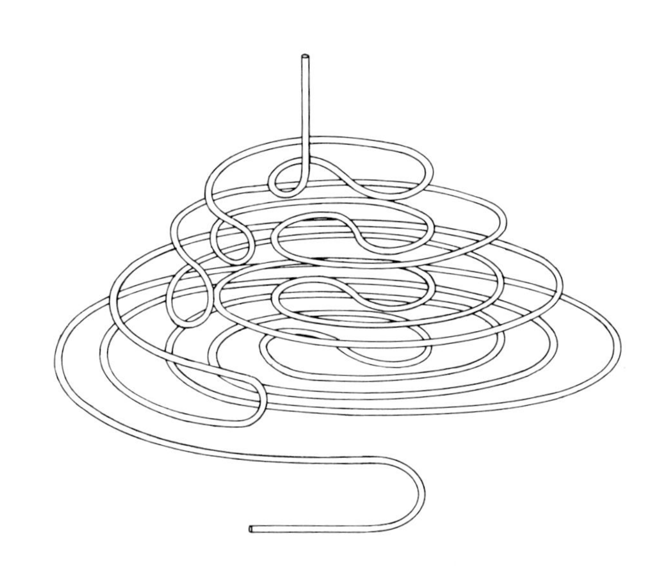

particular fossil species are established, the details of environment, climate, and indeed the history of the life within an ancient landscape, will emerge. The location of the deposit may prove to have been continental in origin—a stream channel, a lake sediment, a desert, or a swamp; or shore related—a delta, an estuary, a marsh, a beach, or a lagoon. Or perhaps it might be shallow marine—a sand bank, an evaporite basin, or a coral reef; or deep marine—a continental slope or submarine canyon. Whatever the origin of the deposit proves to have been, at one time it will have been part of a landscape or seascape which was active, alive, and dynamic. The prevailing climate of the time will have been either hot and arid, warm and humid, or cold and miserable. All these signs are written in the rock strata and can be read like a book.

The area of Grand Canyon, the Painted Desert, Zion, and Bryce canyons probably represents the most extensive region of undisturbed, exposed, and diverse horizontally bedded sedimentary rock on Earth. It is an ideal place to establish stratigraphic boundaries that indicate past changes in environment and climates. Its potential catalog of index fossils provides superb data for biochronology. Here one can establish the life-style of the multitudinous denizens of this particular part of a drifting continent as it moved across the face of the Earth accumulating sediments.

Some of the oldest fossilized animals found in Grand Canyon are "trilobites." They first appear in the Bright Angel Shale of the Cambrian Period on the Tonto Platform. Their habitat, character, and life-style have been studied so extensively and are so well known that they classically typify the usefulness of fossils to geologists.

Trilobites were "arthropods." *Arthropoda* developed into the largest group in the animal kingdom, and perhaps seventy-five percent of the total known species of animals alive today, very nearly a million in all, are arthropods. They are invertebrates with jointed legs (hence *arthro-poda*) and include scorpions, millipedes, and insects.

A feeding burrow is made by an animal that tunnels into sediments of the ocean floor and eats bits of organic matter it extracts from the silt. The burrow also provides shelter. The spiral is a feeding burrow, too—actually a double spiral made by a worm-like animal that circled inward, made a U-turn at the center, and circled outward again in the empty area between its entering whorls (see hypothetical outline above).

The trilobites were a very early group in the *arthropoda* phylum and their proliferate nature undoubtedly influenced the subsequent chain of their evolution.

There are no trilobites living today, nor have there been since Permian times, about two hundred and twenty-five million years ago. The group probably originated in Precambrian times. It is believed to have evolved from the primitive worm-like creatures called "annelids," which have been referred to above. Trilobites literally dominated the shallow Cambrian seas from about five hundred and seventy million years ago. So for a period of about three hundred and fifty million years innumerable species and subspecies developed. Some of these were successful and some were not.

The extinction of any one of the innumerable species of trilobites marks a point in time. The extent of the career of that species represents a definite span of time. From a geologist's point of view, the shorter the career of a particular species, the better. Its uppermost and lowest appearance in a strata or series of strata establishes the time span of that species. In other words, the species can be used as a key or "index" fossil. The irony is that the less successful a particular species of trilobite, or any other animal, the better its qualification as an index fossil, because the time span of its occurrence is narrow and therefore any event associated with it can be more accurately placed in the chronological record.

The name "trilobite" indicates a three-lobed body which consisted of two lobes on either side of an axial lobe. The animal had a head region, called the cephalon, a

thorax, and an abdomen, and each segment of the body bore jointed appendages. Those at the front developed into sensory and feeding organs and those behind were used for breathing through feathery gills at the top of the appendage, and for propulsion at its extremity. Most trilobites had a pair of compound eyes in the head but some had no eyes at all. The whole animal was encased in an exoskeleton made from a protein called chiton which was exuded from the skin—not unlike a human fingernail. As the body grew, the exoskeleton became too small for its owner and so it was shed, perhaps with considerable difficulty, and replaced by fresh chiton which hardened. Many modern arthropods molt from four to thirty times in a lifetime, depending on eventual size. Since trilobites varied in size from a fraction of an inch up to nearly thirty inches (the largest trilobite fossil found in Grand Canyon measures over three inches in length), it is reasonable to suppose that they too molted with about the same frequency as their modern counterparts, the crustaceans. It was mostly the trilobite exoskeletons that were fossilized, so trilobite stages of growth are as well known as their evolutionary development.

Female trilobites laid eggs in small depressions scooped out in the sandy seabed. It is inferred from the behavior of modern crustaceans that these eggs were fertilized by a male trilobite and then covered by loose sand sufficiently well to protect them from predators, often other trilobites. Young trilobites were prolific but just as liable to be consumed by their parents, other trilobites, and predators, as young

Fossils of the Kaibab Sea

The small red fossil shells seen in this close-up panorama are often so perfectly modeled and in such proud relief that instinctively one tries to pick them up as if they are lying on a sandy beach. The beach is itself petrified, and is, in fact, on the North Rim of Grand Canyon at Boysag Point, some 6,000 feet above sea level. 19. E-3.

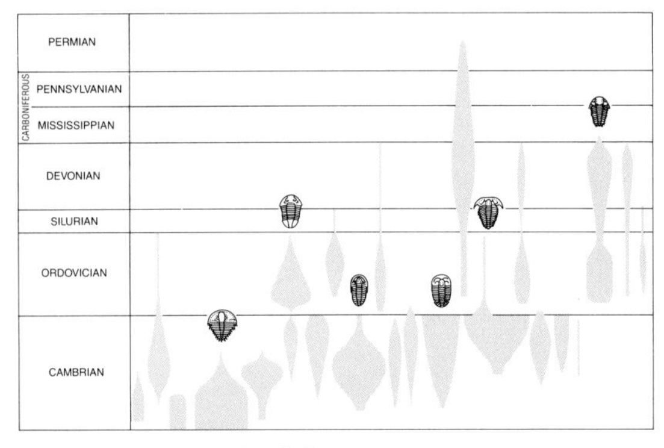

Mass extinction of trilobites

These primitive arthropods died down at the close of the Cambrian period about 500 million years ago. During the Cambrian period hundreds of kinds of trilobites populated the shallow seas of the world. The chart depicts 15 superfamilies of Cambrian trilobites; the width of the shapes is roughly proportional to the number of members in each superfamily. Final extinction took place in the Permian.

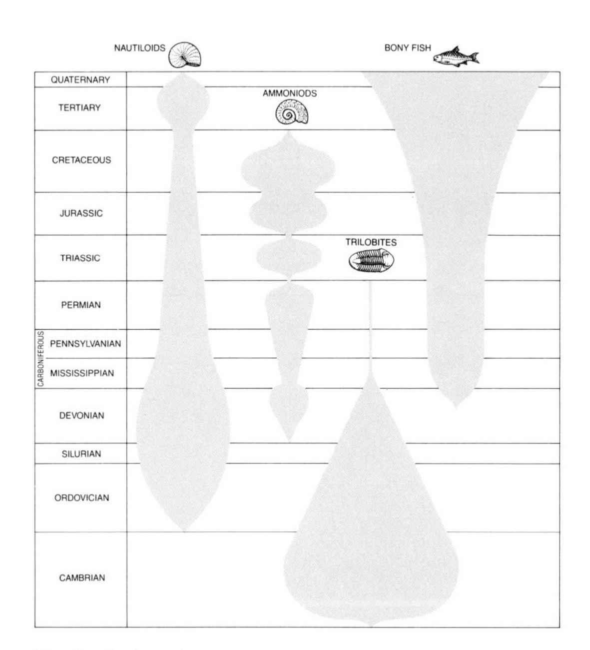

Ecological replacement

Ecological replacement appears to be a characteristic feature of evolution. The diagram shows a waxing and waning among four groups of marine swimmers, dating back to the earliest fossil records. The ammonoid group suffered near-extinction twice before finally expiring.

crustaceans are likely to be consumed after hatching today. If trilobites survived infancy, then depending on the habits of their species, they swam at the surface of the sea or just below it, or near the seabed, or perhaps they just drifted and ate plankton. Some, perhaps the majority, crawled over the sea bottom, some plowed through sand or mud, and others burrowed. Many were scavengers and fed on dead animals and plants and some species in Devonian times learned to curl up into a ball for self-protection like the modern but unrelated pill bug or wood louse.

The mode of life of an individual species determined its anatomical development. Swimmers and drifters tended to develop relatively large eyes so that they had a wide field of vision. Bottom crawlers were the pygmies of the group and developed armor-like shells. Some plowers developed eyes that were raised high on the cheeks of the cephalon so they could see above the mud that covered their bodies, and others developed a spine at the rear end. Burrowers tended to develop small smooth shells and to lose the use of the eyes for which they had little need.

The fossilized shells of trilobite species are similar to the prefabricated steel vehicle bodies made in the automotive industry. Pressed-steel bodies reflect the fundamental anatomical differences between one make and another, a Dodge or a Datsun, and between one function and another, an automobile or a truck. So it is with a trilobite. From the form and position in a rock formation of a trilobite fossil it is possible to deduce whether the animal was an early species or a late one, a swimmer or a burrower, and so on. Trilobite egg pits, seabed tracks, burrows, and even the debris in the immediate vicinity of burrows are trace fossils which enable paleontologists to reconstruct the life-style of trilobites as surely as the tire tracks of a vehicle are clues to its purpose.

After the Cambrian Period, trilobite domination of the seas peaked and began a long decline that ended in their complete extinction at the close of the Permian Period. So, with one glance over the rim of Grand Canyon, one can survey practically the whole three-hundred-and-fifty-million-year episode of this creature, from its first appearance in the rocks of the Tonto Platform in the lower part of the Canyon, to a point in time little short of its extinction, in the Kaibab Limestone at the rims.

Although trilobites dominated the Cambrian seas, the brachiopods (marine invertebrates with a pair of dissimilar half-shells) were a close second to them in numbers. There were also sponges, spiny-shelled animals called *echinoderms*, which eventually evolved into creatures like sea lilies and sea urchins, and sea snails. And there must have been many soft-bodied creatures that left no trace.

Except where the Devonian Rocks lie between the two, a vertical cliff, the Redwall Limestone, lies directly on top of Cambrian rocks with a time-gap of one hundred and fifty million years between. As previously discussed, no Ordovician or Silurian rocks exist in Grand Canyon, but there are a few places in the eastern half where Devonian rocks, in lenses of pale lavender-colored limestone up to thirty feet thick and one hundred feet long, intervene between the Cambrian and the Redwall. Extensive outcrops of Devonian rock appear at the western end of the Canyon where the formation reaches a thickness of over a thousand feet in places.

The Devonian rocks in Grand Canyon contain beds of dolomite, a type of rock in which magnesium carbonate predominates over the more usual calcium carbonate. The dolomite was laid down in very salty, shallow water but the natural process of its formation is an enigma which has yet to be solved. There is evidence to suggest that there were freshwater stream valleys in the eastern Grand Canyon area during Devonian times, and it is in these that one of the early vertebrates, the fossilized fish named *Bothriolepis*, has been found. Although fish first appeared in the Ordovician Period, the Devonian Period is known as the Age of Fish because of the very large number of new species of fish which evolved during this time. *Bothriolepis* was a freshwater animal which grew to a length of about ten inches and frequented streams and lakes. It is of special interest because if it strayed into a tidal estuary and became stranded on a mud flat it was capable of breathing air until water returned.

The great Redwall is the thickest continuously visible rock formation of Grand Canyon. It is Mississippian and aptly named. The Redwall forms a sheer cliff face five to six hundred feet in height that stretches from the eastern end of the Grand Canyon to the western extremity, a direct distance of about one hundred and fifty miles. The true color of this limestone cliff is blue-gray, but it has been stained by the iron oxides washed from the formations of red rock immediately above. The chemical purity of its constituent carbonates has largely determined the physical character of the formation. Erosion has shaped this great red wall into amphitheaters and alcoves separated by soaring buttresses. In part, it has been undermined by caves and riddled with passages. Water through the ages has reacted with it, leaching it, dissolving it, and draining down the steepest incline to sculpture it into monumental curves and columns.

The Redwall was largely deposited in extensive low-energy seas that covered an area stretching a thousand miles north and hundreds of miles westward. During the twenty-five-million-year duration of the Mississippian Period, the sea flooded and receded (the geological expression is transgressed and regressed) from the then coastal area of Grand Canyon three times. The Redwall Limestone is thus a composite sandwich of four layers of deposition. When the seabed sank slowly, a thin layer of limestone was deposited, and when it sank quickly, a relatively thick layer of sediment was laid down. The thickest unit of the formation is more than two hundred feet deep and the other three vary from sixty to ninety feet. Much of the last deposit—now the thinnest—was removed by erosion during the earliest part of the Pennsylvanian Period which followed the Mississippian. There is, therefore, a considerable unconformity between the upper surface of the Redwall and the Pennsylvanian rocks above.

Much of the rock which forms the Redwall consists of an incomprehensibly vast collection of the fossilized skeletons of sea animals which were consolidated into rock. Fish, trilobites, the remains of armored creatures called *ostracodes*, and many more representatives of the primary divisions of the animal kingdom (such a primary division is called a "phylum") are found in this formation. The most common fossils are corals, brachiopods, crinoids—sometimes called sea lilies—and the unicellular foraminifers, which are a fantasy of minuscule design. Life in these seas was literally bursting with energy and diversity. Life on land was developing too, although there is no evidence of it in the Grand Canyon area because the locality was quite evidently a marine environment at that time. But there is ample evidence from many parts of the Earth that the Mississippian Period was a time of warm, humid climate.

The unconformity between the upper part of the Redwall and the overlying

The Court of the Patriarchs (overleaf)

These soaring crags of Zion give an overwhelming sense of height as they rise about 3,000 feet above the Virgin River which carved the valley (hidden by foreground trees). The upper portion of these cliffs was formed from deserts of eolian sand during Jurassic times, about 180 million years ago, during a time of constant searing winds and few signs of life. The Sentinel is at center left, and the Three Patriarchs are at right. 20.D-3.

Pennsylvanian Formation, the Supai Group, looks even and nearly horizontal. Once again distance and scale deceive, for in fact the surface between the Redwall and the Supai represents the last stage of an erosional cycle that extended over a period of perhaps thirty million years during which an interesting and varied landscape was formed. The entire Redwall is very hard limestone and resistant to erosion. Nevertheless, on close examination at the level of the unconformity, it can be seen that it is cut by channels and contains bowl-shaped depressions. These features are interspersed with mesa-like shapes, ridges, ledges, and varieties of undercutting. Some of the depressions are filled, not just with the conglomerates of the lower Supai formation, but with large angular pieces of limestone and other fragments that suggest we are looking at the remains of a collapsed cavern system. The upper Redwall immediately below the unconformity is in fact riddled with caverns which were formed at the same time as the now-eroded and fragmented cavern system which once existed above them.

Before too many new and confusing names are necessarily introduced, the system of applying otherwise mind-numbing nomenclature to the rock formations should be explained. Although rock formations in a locality are assigned to the geologic periods in which they were formed for reference purposes, they are also given their own names. The reason is that the type of rock from which a formation is formed may be totally different from other rock which was formed at the same time in another locality a few or many miles away. For this reason particular formations are allocated place names commonly associated with the locality in which they were first found or are best represented. The etymology of the Supai Group of rock formations provides a good example of the custom.

The best local exposure of the Supai Group is near Havasu Canyon, which is west of Grand Canyon Village on the south side of the Colorado River. "Havasu" is derived from an Indian word that means "blue-green water," a reference to the stream that runs through Havasu Canyon. The Indian people who live in this canyon are called the Havasupai, blue-green-water people. The village in Havasu Canyon in which they live is called Supai, the only Indian village inside Grand Canyon.

The red-bed Supai Group above the Redwall (which in the time sense straddles the late Pennsylvanian and Permian periods) is named after Supai Village. The Group is divided into four formations, each representing different depositional periods and environments. The first three are named after Havasupai families—Watahomogi, Manakatcha, and Wescogami—and the fourth, Esplanade, was the name given by an early explorer-geologist to a broad expanse of inner-plateau into which Havasu Canyon opens and joins the broad prospect of Grand Canyon.

Each formation in the Supai Group contains evidence of at least one sea transgression and regression in addition to distinctive types of environment. Unconformities separate one unit from the next. This is known from the surfaces of erosion and the extensive conglomerates overlying each. The lateral as well as vertical differences in these deposits show that for long periods the vicinity of Grand Canyon was near an ancient shoreline. Marine fossils such as trilobites and brachiopods help determine the boundaries between the sea and the seashore. When the limits of the marine deposits are plotted, a map of the ancient shoreline can be produced.

The oldest of the Supai formations, the Watahomogi, is almost entirely marine.

The small-scale ripple marks and climbing ripples that are present in the strata show that the seas in which deposits were laid down were calm and shallow—low energy. The next formation, the Manakatcha, also mainly marine, was deposited in much deeper and often rough seas. This is deduced from the large-scale cross-bedded type of stratification that is present and that shows there was considerable movement on the seabed at the time of deposition. The next formation, the Wescogami, contains both large-scale cross-bedding and structures that suggest the activity of streams and shallow rivers wandering over a coastal plain. So in this case one can imagine rough seas pounding a continental shoreline, but a shoreline with a difference. Footprints of a four-footed five-toed animal are imprinted in the now-lithified beach, footprints of early amphibious reptiles that ranged and foraged in the area. The Esplanade, the fourth and last of the Supai formations, manifests a mixed environment. The presence of kaolin—china clay—suggests a continental landscape; gypsum—an evaporite deposit—implies that lagoons were once prevalent; corals and other marine animals suggest warm shallow seas to the west; and a particular type of bedding structure in the sandstone indicates a shallow stream and river environment.

The Supai Formation is a formidable and complex deposit of red rock, in some places more than a thousand feet thick. It is surmounted by the Hermit Shale, which ranges from less than one hundred feet in thickness at the eastern end of Grand Canyon to a thousand feet in the western end. As the name implies, this formation is comparatively soft friable rock, and forms a distinctive steep-angled bench immediately below the three- to five-hundred-foot vertical cliffs of white Coconino Sandstone. And above the sandstone is yet another cliff of creamy yellow-gray limestone five to seven hundred feet in height made up of two formations, the Toroweap Formation and Kaibab Limestone, with an unconformity between them.

With the exception of the Coconino Sandstone all these Permian-age rocks were formed in ways similar to, but not the same as, those of the Tonto Group, the Redwall, and the Supai—under the sea or by rivers on the land. They were the result of deposition in low- to high-energy seas that flooded and receded from the now fully formed Pangean continent during periods that experienced a variety of climates ranging from semi-arid to extremely arid. The Coconino Sandstone is an exception because it was formed from eolian sand dunes, which from their structure are judged to have been carried by a wind blowing from an area north of the present position of the formation.

These events of three hundred and fifty million years resulted in the deposition of much more than four thousand feet of sedimentary rock on top of the Vishnu Schists and Grand Canyon Series at the bottom of Grand Canyon. Some of these sedimentary rocks were reduced by erosion, leaving considerable unconformities. But much of the time the rocks that can be seen were under the sea. Each successive formation contains ever-increasing numbers of fossil forms. But at the end of the Permian Period, of which the Kaibab Limestone on the rims of the Canyon is the last representative here, many life forms became extinct. The trilobites disappeared altogether, never to return. The brachiopods, crinoids, and corals suffered a serious decline. It was a time of death and extinction over much of the world, a momentous time in its history when the supercontinent of Pangea had formed, and the dinosaurs first appeared.

All of this is deduced from fossils and stratigraphic boundaries. Perhaps it can now be safely admitted that the heretics Palissy and Bruno were right.

VI

SON OF JUPITER

The San Francisco Mountain volcanic field

This panorama was taken from the summit of O'Leary's Peak (8,925 feet). The highest peak of San Francisco Mountain is Humphrey's Peak (12,680 feet), at the far right. When it was an active volcano it was 3,000 feet higher. San Francisco Mountain started forming about ten million years ago, and continued its sometimes violent eruptions until comparatively recent times. The last violent eruption has been dated to about 4,000 years ago. The last cinder cone to form was Sunset Crater (at center left) about 900 years ago.
21. G-5.

One of the greatest and most successful ventures in exploration was the flight of spacecraft *Voyager I* near planet Jupiter on March 5, 1978. By far the largest planet in the solar system, Jupiter has three hundred times the mass of Earth and eleven times its diameter. At its closest, it is nearly four hundred million miles away from us. Of the fourteen Jupiter moons that have been discovered so far, the four largest were observed by Galileo with the first astronomical telescope early in the sixteenth century, and were named after Roman deities associated with Jupiter. Three of the four were photographed in detail and in brilliant color by *Voyager I* with their parent planet as a backdrop, perhaps the most remarkable pictures ever exposed. Io, one of Jupiter's satellites and a little larger than our own Moon, proved to have active volcanoes that were photographed in eruption—a picture and a discovery of profound importance when contemplating the geophysics and biotic potential of the solar system.

In Roman mythology, the god Jupiter was lord of heaven and bringer of light. Io was a priestess of whom Jupiter became enamored. His wife, Juno, was queen of heaven and patron of married women. When she discovered Jupiter's attachment for one of her minions, he hastily transformed his paramour into a globe-wandering heifer, or so the story goes. Later, Jupiter and Juno had a son called Vulcan who became god of fire and is manifested on Earth by mountains of fire—"volcanoes." The discovery of active volcanoes on the moon Io would certainly have rekindled Juno's suspicions of Jupiter's infidelity. According to legend Vulcan forged his father's thunderbolts in a workshop beneath one of the most persistently active volcanoes in the world, Mount

Etna, Sicily. But the sovereign's throne of Vulcan's violent kingdom is situated in Grand Canyon, Arizona, or at least, an impressive but inactive volcano was so designated in 1869 by Major John Wesley Powell. Powell and his companions, who were the first men to travel the length of the Canyon by the Colorado River, discovered quite a remarkable volcanic system in the western section of the Canyon. The summit of one of its volcanoes is perched on the very edge of the Inner Gorge of Grand Canyon, nearly four thousand feet above the level of the river. "Vulcan's Throne" is one of the many hundreds of inactive volcanoes of the Colorado Plateau, a region which was at one time very much a part of Vulcan's vast and fiery domain.

Volcanoes are a manifestation of Earth's internal pressures relieved by crustal movement and local faulting. There are hundreds of thousands of volcanoes on Earth, about five hundred of which are active. In any one year it is unlikely that more than a few dozen of these will erupt. Occasionally, very occasionally, a dormant volcano will unpredictably come to life with devastating consequences to the local landscape. It is therefore unwise to pronounce a volcano inactive if it has erupted during the last few hundred years. Except in very special circumstances, few geologists would be prepared to tempt fate by calling a volcano "extinct."

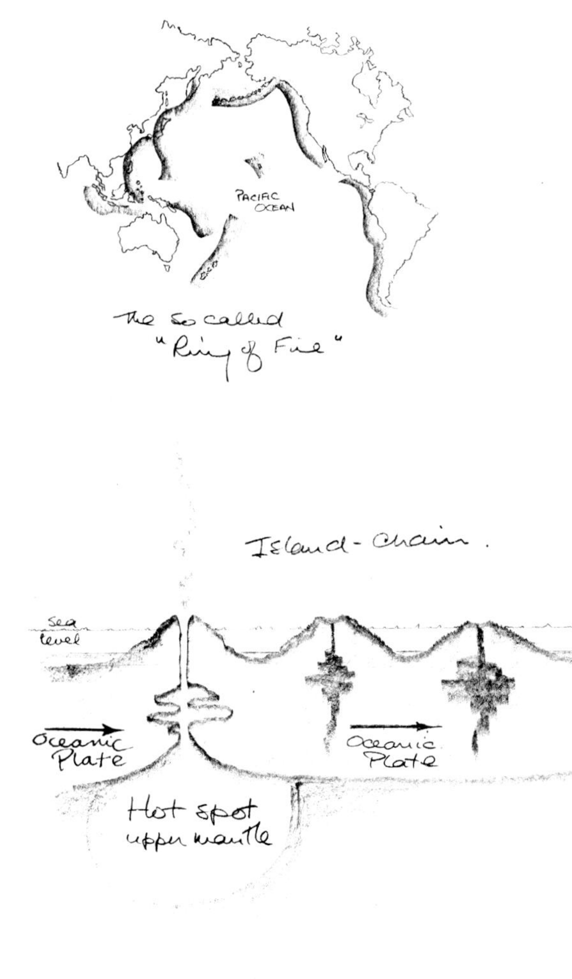

One glance at a map of the Earth's continental plate structure is sufficient to convey that the majority of volcanoes occur within a few hundred miles of the continental plate boundaries (see illustration page 23). It is there, mainly in oceanic and coastal areas, that the crust is least stable and therefore most vulnerable to rupture by the enormous internal pressures exerted from below the surface by the Earth's magma, pressures created by the combined weight of the seven continents and the oceans which float on the lithosphere enveloping the mantle of the planet. So when the Earth's crust is contorted by plate movement, magma percolates into the faults that permeate the plate boundary areas. Such injections cause melting of rocks near the surface, which in turn makes the crust less stable. This interaction can lead to the breach of the upper crust, allowing magma to erupt to the Earth's surface. This concept is vividly illustrated by the "Ring of Fire" of more than ten thousand volcanoes encircling the Pacific continental plate.

There are relatively few volcanic systems away from plate boundary areas. Of these, the Hawaiian Islands are the supreme example of a type called "shield volcanoes." It is thought that the Pacific oceanic plate moves across a permanent hot spot in the Earth's mantle, which acts like a fixed welding arc under a moving sheet of metal. As the plate moves, one volcano ceases to function and a new one starts up, leaving a string of truly extinct volcanic islands behind. The Hawaiian group includes the tallest volcano on Earth, Mauna Loa, which when measured from the Pacific Ocean bed to its highest point above the sea—the island of Hawaii—is about as high as Mount Everest.

Some continental volcanoes near to plate boundaries are the by-product of subduction which produces "hot spots" inland, where one plate is overriding another. Other inland volcanic systems, well removed from apparent subductive influence, are considered to be the products of regional movement of the Earth's crust caused by our old friends orogeny and isostasy—mountain building and surface level compensation—the Tweedledum and Tweedledee of geological happenings.

The main element of a volcano is a vent through which magma either flows or is ejected with varying degrees of force. The temperature and the composition of the magma, and the gases which combine with it, determine the consistency of the lava, which is called its "viscosity." The composition of lava determines both the type and behavior of a volcano. Shield volcanoes of the Hawaiian type are mainly built up from very hot basalt of "liquid" consistency, and therefore of low viscosity with negligible gas content. The volcanoes which result have a low profile. They are unlikely to explode. The conical type of volcano—termed "strato-volcano"—is generated by a mix of viscous lava flows formed from magma with a high silica content and with a high gas content; such magmas are ejected explosively to form bombs or cinders or ash or mixtures of all three. Composite strato-volcanoes can occasionally and quite literally blow their tops. For example, Vesuvius erupted explosively to bury Pompeii and Herculaneum in A.D. 79. But the classic example is the annihilation of the volcanic island of Krakatoa near Java in 1883 which, after two hundred years of dormancy, exploded with a force almost seven times that of the Hiroshima bomb, killing everyone in the vicinity and thirty-six thousand others on neighboring islands.

Flood basalts are lavas emitted from fissures rather than from central vents, but they frequently have associated cinder cones. Basalts of this type are so fluid that successive flows build up considerable thicknesses on neighboring horizontal plains, sometimes to a thickness of hundreds of feet. On the other hand, explosive, "pyroclastic" lavas can be emitted from either fissures or vents. Such lavas are very fluid while charged with gas in the magma chamber, but can become extremely viscous when the gas has been discharged.

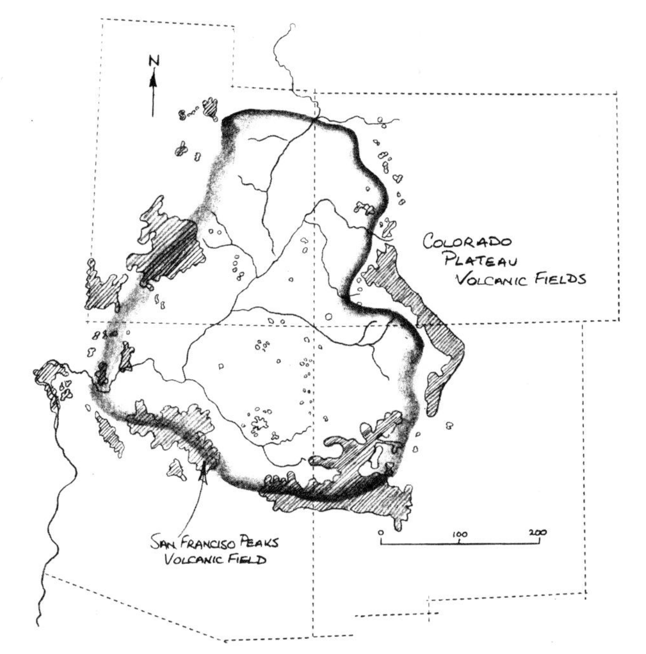

The many (presently dormant) volcanic fields of the Colorado Plateau developed during faulting in the Earth's crust caused by the relief of stress during uplift. Such faults sometimes allow magma from the asthenosphere to escape on to the Earth's surface, and in fact some of the Plateau's principal volcanic fields coincide with its fault boundaries. The whole region may be thought of as a series of slightly elevated blocks with their edges coincident with regions of seismic activity but with little such activity under the Plateau itself. Earthquakes are the physical expression of movement along faults and a major fault movement can cause a major earthquake. The measurable movement which has taken place in some Colorado Plateau boundary areas in recent times suggests that it is still moving sporadically but appreciably. Local volcanoes may not suddenly burgeon into violent eruption in this century or in the next, but the time of volcanic activity on the Plateau probably has not ended.

The vulcanism in the Grand Canyon, Zion, and Bryce Canyon areas coincides with specific fault zones. But by far the most comprehensive example of volcanic activity in the Plateau is San Francisco Mountain of northern Arizona, which dominates the southeastern approach to Grand Canyon. This takes the form of a cluster of high summits, snow-clad through most of the year, soaring to a height of 12,670 ft. above sea level. This elevation is more than five thousand feet above the general height of the surrounding landscape, which slopes down toward the Little Colorado River and the Painted Desert twenty miles to the east and up to the South Rim of Grand Canyon sixty-five miles to the north.

Because of the relatively featureless surrounding terrain, the triple summits of the San Francisco Peaks can be seen from a distance of well over a hundred miles. In recent geological history the trio were one, a single cone of an active composite volcano of

Anatomy of a Volcano

There are a number of extensive volcanic fields in the region of Grand Canyon and Zion. Volcanoes are also associated with the Bryce Canyon area, but to a much lesser extent. There are some four hundred cinder cones in the San Francisco Mountain field, and many hundreds more in the Uinkaret and Shivwits plateaus at the western end of Grand Canyon. When this number is added to those of Zion, the total must be in excess of a thousand. This illustration depicts the shapes of magma that form under volcanic regions. In the diagram, magma has intruded the Earth's crust by sheer force, disintegrating and melting it and pushing it aside. Disintegrated rocks ("xenoliths") are chemically digested by the magma. Sedimentary rocks are metamorphosed by contact heat. The Redwall Limestone, for example, is thereby converted to marble. "Dikes" are near-vertical sheet-like intrusions which can be up to hundreds of feet thick; when they run in parallel they are called "dike swarms"; as shown, they can also take the form of "ring dikes" and "radial dikes." "Sills" are magma intrusions that follow the lateral bedding of rock formations, which in the Grand Canyon area are nearly always horizontal. Viscous lavas may form sills with blisters that force up overlying rocks and are called "laccoliths." A "stock" or "irregular boss" is a mass of intrusive rock with a craggy surface; it is roughly circular, with steep sides. Many of these features are revealed by the erosion of the covering surfaces. When a volcano is stripped of its outer layers of cinder and lava flows, the now-solid inner core or "pipe" is revealed. This is called a volcanic "neck." Huge masses of granite known as "plutons" may accumulate underground, and sometimes extend for hundreds of square miles as "batholiths." Plutons and batholiths can also be raised to the surface by uplift and stripped by erosion to form part of the landscape.

the Vesuvian type, three thousand feet higher than the present eroded level of the highest peak. The volcanic field, of which San Francisco Mountain is still the central feature, is about twenty-two hundred square miles in extent. The field contains several major volcanoes in addition to San Francisco Mountain and many hundreds of cinder cones varying in height from a few hundred to a thousand feet above the local countryside. The most recent volcano to erupt there was Sunset Crater in A.D. 1066, an event which had a profound effect on the people who lived in its vicinity at the time—a story for a later chapter.

The ten-million-year volcanic history of the San Francisco field began during the late part of the Tertiary Period of the present Cenozoic Era after an extended period of erosion had stripped the area down to the level of the Shinarump conglomerate, which then formed the surface rocks of this part of the Plateau. These rocks lay on top of the Moenkopi Formation, which in turn surmounted the Kaibab Limestone—the present surface. At this point it is probable that violent earthquake tremors opened up a series of vents and fissures which allowed the first basalts to flood the area.

During this extended period of vulcanism, which lasted about six million years, three thousand square miles of this part of northern Arizona were covered by about thirty

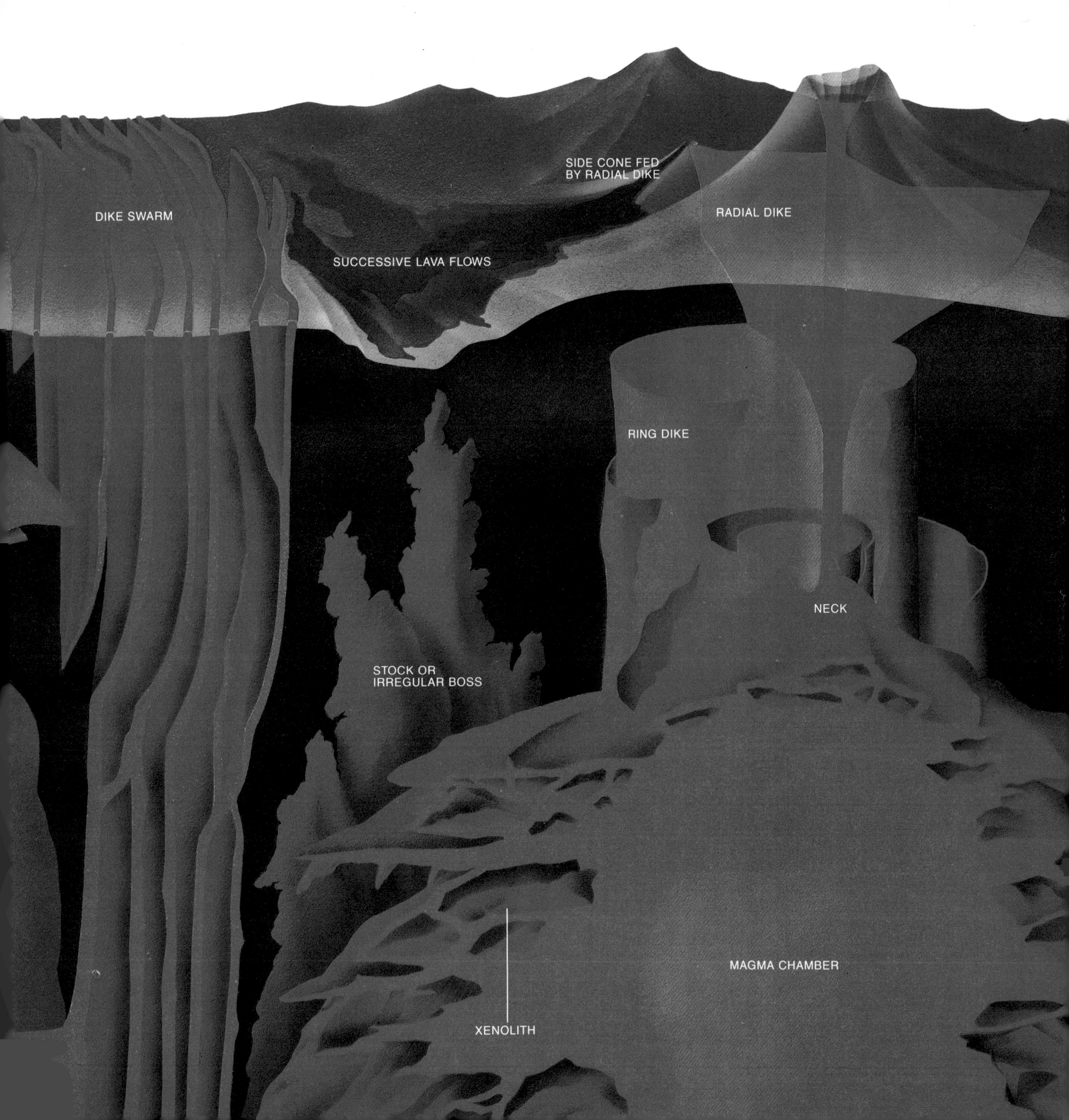

SIDE CONE FED BY RADIAL DIKE
DIKE SWARM
RADIAL DIKE
SUCCESSIVE LAVA FLOWS
RING DIKE
NECK
STOCK OR IRREGULAR BOSS
MAGMA CHAMBER
XENOLITH

cubic miles of basalts. The floodplain cover was between twenty-five and seventy-five feet thick. Occasional depressions in the landscape and streambeds which crossed them filled with basalt to a depth of several hundred feet. The basalt also formed a platform fifteen hundred feet thick on which the present San Francisco Peaks stand.

Radiometric dating in the area indicates that a period of three million years followed during which there was comparatively little volcanic activity. But the Plateau continued to uplift, and during this same period seven hundred feet of the Shinarump and Moenkopi formations were largely eroded away, leaving the area much as it is today. There was no San Francisco Mountain yet, only the basalt platform, a number of small cinder cones, and two buttes capped by lava (now called Red Butte and Cedar Butte) that had been protected from erosion.

The second period of volcanic activity, which began about a million years ago, was entirely different from the first. It was extremely violent. Basalt was no longer the main constituent; the magma now had a high silica content that produced many types of lava that were extremely fluid when charged with gas in the magma chamber under high pressure. But the moment they were forced through a vent, the pressure was reduced, the gas expanded, and the viscosity of the lava increased exponentially. Such highly explosive mixes caused violent eruptions. It has been estimated that there were up to two hundred periods of eruption in five prolonged stages, and that two-thirds of these were violent. It was at this time that San Francisco Mountain and neighboring major volcanoes were formed, including five major and ten minor cones. The total volume of magma ejected to form San Francisco Mountain alone was about thirty-eight cubic miles, of which approximately three cubic miles have been eroded. The scene must have been similar to Vulcan's home territory when Mount Etna is in full spate.

The anatomy of a volcanic region similar in character to San Francisco Mountain is generally described in the illustration on pages 100–101. One of the features portrayed is a *laccolith*—a volcanic intrusion caused by the injection of magma between the horizontal planes of rock formations below ground. Laccoliths were formed in the San Francisco Mountain region in this second general period of eruption, while the main volcanoes were being shaped. One of these was under such great pressure and was so vigorously intrusive that, after penetrating the Redwall Limestone about two thousand feet below the surface and turning parts of it into marble by contact metamorphism, it burst through the remaining formations to create a lava mass now called Elden Mountain, three thousand feet high.

There was no discernible pause between the second and third periods of eruption. They are distinguished by the completely different character of their lava flows rather than by time interval. Silicic lava flows ceased, and basaltic flows similar to those of the first period once again predominated. These erupted vigorously and frequently at first, but gradually subsided until they ceased to flow, less than a thousand years ago. During this last period of general eruption, twenty cubic miles of lava were spewed over the landscape, much of it superimposed on older flows. Several hundred new cinder cones were also formed on the western and eastern approaches to the Peaks, Sunset Crater being the last.

Immediately to the east of the San Francisco Peaks the volcanic field slopes down toward the Little Colorado River, which at this point had cut a canyon one

Reclaiming a canyon

Grand Falls formed after the Little Colorado River Canyon was blocked by a basaltic lava flow from a nearby volcanic vent. The river made its way around the tongue of lava and back over the Canyon wall into its original channel. The heavily sedimented river is the recipient of the spring runoff from the high country surrounding the Painted Desert. The absence of distortion makes it difficult to appreciate that as one faces the Falls from where this panorama was taken, the San Francisco Peaks at top left are in fact behind the viewer. To see them one would have to turn one's head. Another panorama of Grand Falls may be seen on pages 34-35.
22. F, G-5.

The Lava Cascades of Toroweap (overleaf)
Here, at the western end of Grand Canyon, the cliffs rise about 3,000 feet from the river. Vulcan's Throne, at the top left of the panorama, was one of the last of a series of volcanic events that began a million years ago. Vulcan's Throne surmounts a series of flows which have filled the mouth of a valley and possibly its full length for many miles back from the Colorado River on the north (left) side. Succeeding flows of thick lava blocked the river itself and formed a high dam. Lava Falls—considered by many to be the most severe whitewater on the Colorado—can be seen at the foot of the Lava Cascades. Upriver from the rapid a sixty-foot black basalt volcanic neck, called Vulcan's Anvil, can be seen protruding from midstream.
23. E-3.

hundred and twenty-five feet deep through the upper part of the Kaibab Limestone. During the last stage of basaltic emission, several lava streams flowed down into this canyon. One of them followed the course of a wash and completely filled the canyon for a distance of one and a half miles downstream. It not only filled the canyon to its rims, but spilled over to cover an area of the east bank beyond. The river at first backed upstream to form a small lake, which then overflowed to create a new course around the lava tongue. Having circumnavigated the obstruction, the river rejoined its old course downstream by flowing back over the edge of the canyon's east rim (see panoramas pages 34–35 and 102–103).

During the few thousand years which have passed since the canyon was blocked, the river has completely removed the lava at the foot of the obstruction and has eroded the edge of the original east rim into a series of steps. The steps seem to represent stages in the sedimentation of the rock formation into which they are cut—some strata appear to be less resistant to water erosion than others. And so it is that a new canyon is being cut to replace the section that was filled with basalt. The heavily sedimented Little Colorado now cascades over the east rim after pouring down the series of steps, four hundred feet wide, before plunging more than a hundred feet into its old channel. Witnessing the Grand Falls of the Little Colorado in full mud-laden flood, thundering over the east rim of the old canyon, is one of the most interesting experiences of the Plateau country. The whole scenario is a microcosm of the mainstream of the Colorado River as it must once have been before it was emasculated by man-made dams. It is a reconstruction in miniature of the forces that formed Grand Canyon and the reaction of rock formations to them. Just to stand and stare at Grand Falls, to watch and listen as it tears at the limestone cliff and pounds at the basalt dam, is to witness the drama of the irresistible meeting the immutable.

There was once a greater lava dam that blocked a greater river, and this dam was not a hundred, but certainly eight hundred feet high, and possibly very much higher. One recent study even suggests fourteen hundred feet. And this was only one of a series of lava dams that were formed in the same vicinity. For a while each successive dam held back the mighty Colorado River. The first of these lava dams was formed just over a million years ago at the foot of Vulcan's Throne. This volcano stands on the line of the Toroweap Fault which crosses the river at a point in the western region of Grand Canyon, one hundred and eighty river miles below Lees Ferry.

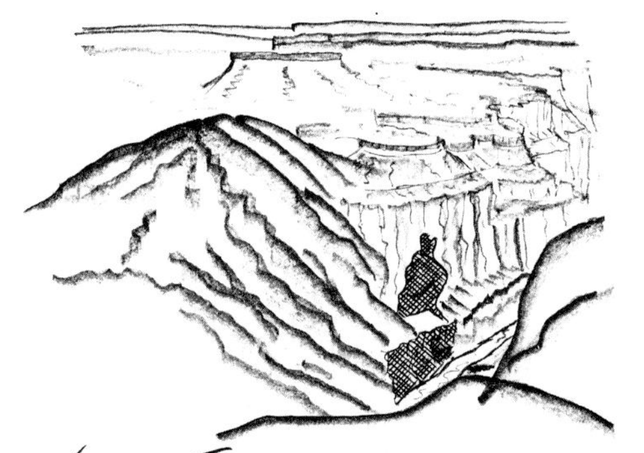

Vulcan's Throne at Toroweap – the lava dam was certainly the height of the lower cross-hatch area and possibly that of the higher one.

There is something magical about the whole locality of Toroweap. It is remote. It is difficult of access. It is starkly spectacular. Yet somehow, in some way, it is a personal place for those who experience it. One feels that one is in an inner sanctum of Grand Canyon. Toroweap's three-thousand-foot towering, tiered cliffs soar majestically from both sides of the Colorado River straight up to the Esplanade above. There are volcanoes all around, and now-solidified lava cascades down to the river. At river level there are lava dikes, volcanic necks, cliffs of columnar basalt with river gravels on top, and terraces of lava precariously perched like hanging glaciers a thousand feet above one's head. Toroweap defies analysis and almost defeats description, but Powell got the spirit of it when he wrote of his discovery on August 25, 1869: *Great quantities of lava are seen on either side; and then we come to an abrupt cataract. Just over the fall a cinder cone, or extinct volcano, stands on the very brink of the canyon. What a conflict of water and fire there must have been here! Just imagine a river of molten rock running down into a river of melted snow. What a seething and boiling of the waters, what clouds of steam rolled into the heavens!*

The Toroweap Fault really created the scene. It is a "normal" fault (see illustration page 43) that was caused by a sizable movement in the Earth's crust. An area to the west of the fault line dropped in several stages a total of five hundred feet almost at right angles to the present line of flow of the Colorado River. The unrelenting uplift of the Plateau was the prime cause. At the time there undoubtedly was a series of shattering earthquakes in the region which resulted from fault movement. And Toroweap was only one of several fault zones in the vicinity, of which Whitmore Wash on the Hurricane Fault a few miles west was the nearest.

The Toroweap Fault produced a line of weakness more than a hundred miles long on both sides of the river, causing valleys to be excavated by rain and snowmelt. These were formed into tributary canyons of Grand Canyon—Toroweap Valley on the northern side, and Prospect Canyon on the southern side. The first of a series of volcanic eruptions began between one and one and a quarter million years ago.

The lava flows blocked the river and created a dam certainly six hundred feet high, and possibly very much higher. The base of that dam on the river bottom at that time was only fifty feet higher than the present river bottom, which is especially significant because it means that there has been little downcutting of Grand Canyon during the last million years. The height of this and other lava dams at the same site can be determined by matching the height of corresponding levels of lava remnants found clinging to the cliffs on both sides of the river today. The lava also flowed downstream, but unlike the Grand Falls episode, there was no way for the river to go around the lava flow because the walls of Grand Canyon were too high. The river simply rose six hundred feet and then raged down miles of solidified and semisolid basalt. The cliffs on the south side, a mile or so downriver, were deeply eroded, and eventually undercut to form a new river channel. The abandoned section of canyon gradually filled with basalt, volcanic ash, cinders, and rock debris, and was finally capped with a small cinder cone, like a pyramid over a sepulcher.

The river water backed up above the dam and formed a lake. The actual size of the lake depended on the extent of breaching by the river as the dam formed. A dam six hundred feet high with a perfectly horizontal top from one side of the Canyon to the other would have backed upriver for about one hundred miles. But because the lip of the dam must have been extremely rough, with sluices pouring through several parts of it, the river is unlikely to have backed up more than part of this distance. Evidence of such "ponding" has so far been limited to that found at Kanab Creek, about forty miles upstream. But one thing is certain: The ride over that dam in a boat would have been even more exciting than running the formidable Lava Falls of today, which are in the vicinity of the old lava dam.

There were later and equally prolific lava flows from Toroweap Valley that covered the original deposits. The lavas that had formed the major dams were thick basalts and were characterized by the long basaltic columns that festoon the exposed cliffs below Toroweap today. The flood basalts which followed were much more fluid and probably traveled a considerable distance along the already partially filled valley. It appears that the valley was filled with lava to a depth of two thousand feet at its foot

Lava dam at Toroweap (overleaf)

A few days before Major John Wesley Powell and his companions completed their momentous first traverse of Grand Canyon in August 1896, they reached the Toroweap region with its impressive cascades of lava above the river. Surmounting the scene was a volcano which Powell named "Vulcan's Throne." He was greatly impressed by the apparent fact that the lava flows had once blocked the river—for there were remnants high above both banks. In fact, geologists believe that Vulcan's Throne was not a contributor to the dam which has been reconstructed on this panorama, but was probably the most recent of many volcanic events in this area. At one time there was a deep canyon running to the left of the picture, and this is coincident with a fault line, known as the Toroweap Fault, which crosses the river at this point. The total displacement caused by the fault was approximately 800 feet in two stages. The total extent of the fault can be judged by the lower level of the rock formations on the left as compared with the same formations on the other side of the Canyon. Advantage has been taken of this difference in height by showing a vertical section through the river, coincident with the fault line. The dam has been drawn at a height of 600 feet above the present river level—about the same water level as Lake Powell and Lake Mead at their respective dams. Some geologists, including Powell, believe the dam could have been twice that height. Toroweap Valley is now completely filled to its end. *(See panorama pages 104–105.)* *24. E-3.*

and that the flood lavas then spilled over the lava cliff at this point to form the Lava Cascades (see panorama pages 104–105 and illustration pages 108–109).

A later series of fault–earthquake sequences caused a further drop in several stages of one hundred and fifty feet downriver from Toroweap Fault. Vulcan's Throne formed its six-hundred-foot cone at the mouth of Toroweap Valley at about this time, surmounting the accumulation of lavas, two thousand feet thick, that had filled and buried the valley. The volcano contributed ash and cinders to the display of pyrotechnics and more lava flows to end the long program of Toroweap's vulcanism, bestowing a final application to the cascade down to the Colorado River. Vulcan now sleeps a deep sleep on his throne, but some listen expectantly for the rumble of movement each day, and live for the time when the master will perform again.

At about the same period of time, Whitmore Wash, on the Hurricane Fault nine miles downstream from Toroweap Valley, was also filled with basaltic lava flows. Between the two places and for many miles back from the rim of Grand Canyon, there is a multitude of now-dormant cinder cones and volcanoes. These are dominated by Mount Trumbull (8,028 ft.) on top of the Uinkaret Plateau to the north and Mount Dellenbaugh on the Shivwits Plateau to the west. After the time of the first lava dams, flood basalts from this volcanic field overflowed the north rim of the inner gorge of Grand Canyon between Toroweap and Whitmore and flowed down the Colorado River riverbed for a distance of more than eighty miles. Since then the river has cut its way through these beds to its present level. This has left boulders, pebbles, and other evidence of its old course collectively termed "hanging river gravels," lying high above on top of the old basalt beds, often three hundred feet thick.

Vulcan's Forge
— a volcanic neck protruding from the river at Toroweap.

Like its counterpart north of the river, Prospect Canyon on the Toroweap Fault plane on the south side of the river is filled with lava to a depth of two thousand feet. But there is a difference between Toroweap and Prospect. While the Toroweap Valley fill at the Canyon's mouth represents an accumulation of thin lava flows over a foundation of thick, the Prospect Canyon fill appears to be entirely built up from thick basalt flows believed to be from the first generation of volcanoes in the region. Also, unlike Toroweap Valley and Whitmore Wash, erosion has cut back quite severely into the lava fill at the head of Prospect Canyon, forming a deep V-shaped amphitheater of vertical basalt cliffs. While making things difficult for close inspection, erosion of the basalt cliff has at the same time exposed an extraordinary cross section of an old volcano which was once buried under the cinders and lava flows of succeeding volcanoes. Although hidden from view, it is obvious from the position of the volcano that the magma vent that caused its formation had to penetrate six hundred feet of Redwall Limestone before bursting forth.

Erosion in Prospect Canyon in the vicinity of Toroweap has also exposed huge dikes of basalt, in one case between thirty and forty feet wide, which have been stripped of once-adjacent rock so they now stand like sculptured ebony walls drawn from a mold. Similarly, but this time rather ostentatiously protruding from the very center of the river above Lava Falls, there is another product of a volcanic eruption called "Vulcan's Forge." At one time this huge block of basalt was thought to have fallen from the cliffs above, but a recent study has concluded that it is a volcanic neck—the now-solid core of a volcano that erupted through a vent in the riverbed. This unique monument to Toroweap's vulcanism towers some sixty feet above the surface of the river and is about seventy feet in diameter. The basalt columns from which it is formed radiate from the central part of the mass and are of a character different from nearest lavas found on canyon walls.

These volcanic souvenirs of the events at Toroweap are far from being antique in the geological sense. In fact some of the oldest known examples of volcanic activity in Grand Canyon are more than one thousand million years old. They appear in the formation of the Grand Canyon Series of Precambrian rocks, which are comprehensively exposed in the Unkar Creek region at the eastern end of Grand Canyon. More than a hundred miles of Granite Gorge separates Toroweap from Unkar, a hundred miles of some of the oldest rocks in Grand Canyon. But at the top end of the gorge at Hance Rapid, five miles below Unkar Creek, a section of the Grand Canyon Series is exposed above the river. This is the site of a basalt dike that penetrated the Precambrian sedimentary rocks an incredibly long time ago. The dike at Hance, seventy-five feet wide, is an impressive volcanic intrusion cutting through the Hakatai Shale of the Grand Canyon Series at a sharp angle. The injection did not have sufficient pressure behind it to penetrate the more resistant Shinumo Quartzite and Dox Formation which overlay the shale, so it permeated the surfaces between the Hakatai and the Shinumo to form sills. Samples of this very ancient dike were carefully selected for the exacting process of radiometric dating. The results established that the dike was formed about one thousand one hundred million years ago.

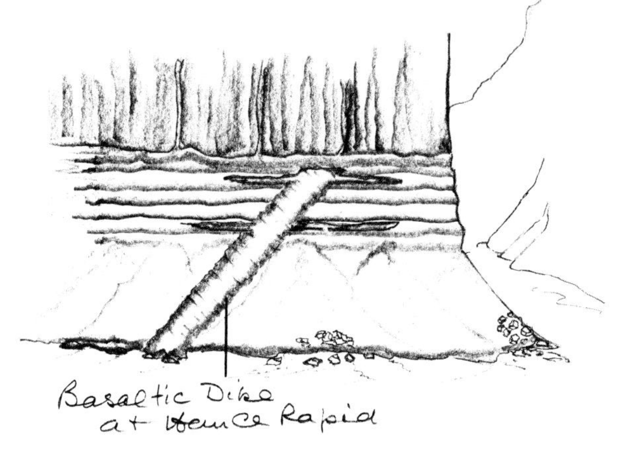
Basaltic Dike at Hance Rapid

This was a very important date to establish. The dike intrudes into some of the oldest sedimentary rock in Grand Canyon. Since the dike penetrates sedimentary rock considerably older than itself, the dating establishes an upper bracket of age for these rocks. Beneath the Hakatai Shale is another and final formation called the Bass Limestone which is the lowest and therefore the oldest of all the sedimentary rocks in Grand Canyon, and this is exposed a short distance downriver from Hance Rapid. Bass Limestone lies unconformably on the Vishnu Schist, which has been established to be about one thousand seven hundred million years old. Because the Hakatai, Shinumo, and Dox formations overlaid the Bass Limestone before the volcanic dike was intruded, it can be reasonably assumed that the Bass Limestone was formed between twelve and sixteen hundred million years ago.

It so happens that the oldest form of life positively identified in Grand Canyon has been found in the Bass Limestone. It contains traces of blue-green algae. These unicellular forms of life are a kind of ancient seaweed called "stromatolites," which were formed in the limey sediments of a shallow primordial sea near the ancestral continent that preceded Pangea. Because of the dating of the basalt dike intrusion, the time of that early form of life can be reasonably assessed. It was probably about one thousand four hundred million years ago.

The unpredictable, tempestuous, and frequently violent phenomena that the Romans dubbed "volcanoes" produced the primordial atmosphere. They were essential to the physiochemical processes that triggered the beginning of life on Earth. Without volcanoes Earth might have been lifeless. And what about the active volcanoes on Jupiter's moon—do they imply life? The answer must be the same as the one given to the question about the tarantula wasp and human vulnerability. The wasp has the apparatus. So has Io.

VII
SPECTRUM OF LIFE

Dana Butte

Many days of every year for millions of years Dana Butte below the South Rim at Hopi Point has caught the last rays of the sun moments before the Canyon is plunged into brief twilight, then darkness. For the author, Dana Butte symbolizes the mystery and the timelessness of Grand Canyon—perhaps a fitting illustration to introduce the prospect of life on Earth which is the subject of this chapter.
25. F-4.

On another planet in our galaxy an exobiologist was examining the latest batch of high-resolution videotape transmissions of planet Earth received from a space vehicle. Previous probes had helped determine that physiochemical conditions for life existed on the planet and that the probability for life's development there was high. But if life had actually developed, how far had it advanced and what form had it taken?

Surface features in certain continental regions of the planet implied the existence of complex arterial pathways. Increased magnification of the tape's imagery indicated movement along the pathways that, though frenzied at arterial intersections and terminals, behaved with random logic from point to point. This was exciting new evidence for life. It suggested communication between living organisms. Video resolution was fine-tuned and magnification increased to maximum. Now there was no doubt, the objects were animated. But were they alive? Did they breathe, eat, excrete, and procreate? And if they did, were they intelligent?

Careful analysis of the videotape evidence resulted in the conclusion that there appeared to be an advanced form of life on Earth. The creatures that had been observed on the planet breathed the atmosphere that consisted mainly of nitrogen and oxygen. They consumed and metabolized hydrocarbons and excreted a mixture of carbon monoxide and water vapor. They seemed to have developed a cloning system of reproduction. Well-organized community behavior indicated a degree of intelligence. The exobiologist noted one most unusual and puzzling detail: The inhabitants of Earth appeared to have wheels in place of arms and legs.

The point of the story is simply to show how easy it is to form the wrong conclusions from apparently reliable observations. Before making any sort of absolute statement, that exobiologist from another world would have wanted at least to land a machine on Earth to perform experiments similar to those conducted by *Viking Lander* on Mars in 1976. And if a carbon copy of *Viking* were to arrive here from a distant planet, what better first place could there be for it to land than on Escalante Butte in the Grand Canyon of the Colorado? A shovelful of Supai Group surface debris and a panorama of that view would give our distant friend positive evidence for life on Earth and allow conclusions to be drawn about its atmosphere, climates, and geology (see panorama pages 15–18).

Although a great deal is known about "life," no one has yet satisfactorily defined the word in terms acceptable to scientists. But in order to research the processes by which life began on Earth, there obviously needs to be an understanding of the difference between matter that is or has been alive, and matter that is neither. Over many years a series of definitive characteristics of life have gradually been validated and found acceptable by most biologists.

The first requirement for life is the presence of the element carbon in its molecular structure. Such substances are called "organic." The second is whether or not the organic material is or has been capable of using, or "metabolizing," externally-derived materials to sustain itself and to produce energy. The next essentials are the ability to reproduce and to grow. The fourth requirement is that the reproduction process must be just less than perfect so that variations that allow adaptation to environment can take place; in other words, that a species can "mutate." Finally, the organism must either positively or negatively respond to changes in its environment, a reaction termed "sensitivity."

These predilections can be put into one brief statement. Living organisms are discrete combinations of self-reproductive, organic molecules capable of mutation and growth, which can metabolize external energy resources and respond to changes in environment. This suggests what living organisms are but does not explain what motivates organic molecules to start the process of "living," or what enables them to interact once they are alive. The trigger of motivation is an imponderable, and considered by many scientists to be a matter for philosophical or metaphysical speculation. On the other hand the mechanism of interaction is well understood.

Conditions on Earth during its formation were extremely inhospitable. There can be little doubt in anyone's mind that a semi-molten surface at a temperature not far short of a blast furnace was incompatible with the existence of living organisms. But the instant that water began to condense in the atmosphere above the surface, the possibility that some form of life would develop increased. When Earth's surface cooled sufficiently to permit pools of near-boiling water to form, that possibility became a probability. It is quite an extraordinary fact that this momentous event in the history of life on Earth, which occurred about four thousand million years ago, is frequently recreated in school laboratories when students routinely repeat the classic experiment first conducted by Stanley L. Miller and Harold C. Urey at the University of Chicago in 1952.

The basis of the first Miller and Urey experiment was indirectly suggested by the work of Aleksandr Ivanovitch Oparin, a renowned and revered Soviet biochemist. Oparin proposed that the first organisms on Earth had derived their basic constituents from the primordial environment. Miller and Urey reproduced their idea of such conditions by putting distilled and sterile water (H_20) into a sealed flask from which air had been evacuated and replaced with a mixture of methane (CH_4), ammonia (NH_3), and hydrogen. The water was kept boiling for a week while the mixture of water vapor and gases was recycled in the presence of a powerful electrical discharge. By the end of the first day the boiling water had become cloudy and pink. The intensity of coloration increased until after seven days it was deep red.

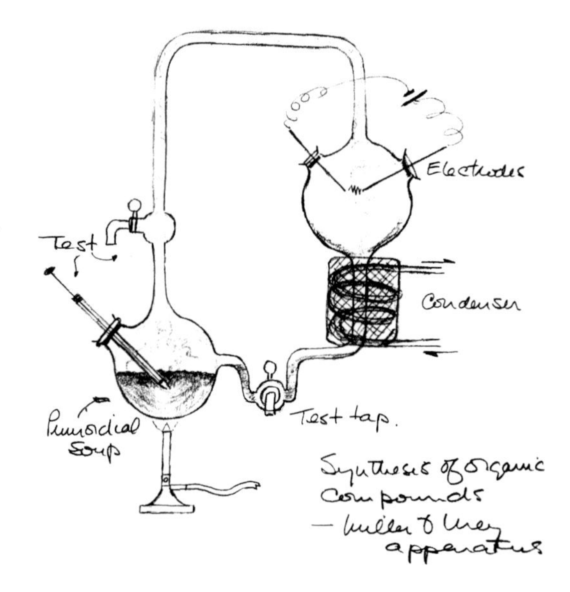

Although the experiment began with simple substances that have very limited powers of combination, it had ended with complex substances that have molecules of unlimited potential for molecular growth. Nineteen such complex organic substances were identified. The work of separating and identifying organic compounds is an arduous task, but could never have been more rewarding than on this occasion. Two types of amino acids proved to be present, and amino acids are the building blocks of natural proteins. Nine other types of organic acid were separated, all essential to life in a variety of ways, perhaps best exemplified in this case by lactic acid, which plays an important part in the metabolism of sugar. Among the remaining compounds, there proved to be a mixture of comparatively simple substances which are termed "polymeric" compounds because they have the property of combining in chemical reactions to form very large and complex organic molecules. What a moment of discovery all this represents—no less than the creation of the primary materials of life on Earth from four very simple chemicals and an electric spark, produced in a glass bottle over a Bunsen burner in a corner of a laboratory.

After Miller and Urey had published their first paper, early in 1953, they and many biochemists in different parts of the world continued to explore the nature of the primordial soup—for it was indeed a "primordial soup" that Miller and Urey had produced. By using a variety of inorganic chemical ingredients, different energy sources, various temperature and pressure gradients, but fundamentally the same technique, experimenters produced an extensive range of hydrocarbons, fatty acids, and amino acids. Most important of all, they created the radical constituents of nucleic acids. These "base" substances combine with phosphorus and sugar to form deoxyribonucleic acid. This is the primary genetic material of all known living organisms, and is more commonly known as DNA.

All that was needed to produce this wide range of key substances was a simple "atmosphere" of gases that contained the elements of carbon, nitrogen, and hydrogen, plus water vapor and an input of energy. It seemed that the energy input could be any one or a mixture of the physical phenomena of the primordial Earth: heat, lightning, ultraviolet light, gamma and cosmic rays. Even shockwaves, which reproduced the effect of the asteroid bombardment of Earth, were productive. Ultra-high temperatures were reproduced in the reaction chamber in the presence of metallic ores to recreate primitive volcanic conditions, and these too were organically productive. So now there was incontrovertible proof of Oparin's theory that given the right physical conditions the chemical building blocks of life are synthesized spontaneously.

The phrase "building block" is apt. Separate molecules of these materials join together in strings, circles, or spirals to form more complex substances. For example

Kwagunt Butte from Chuar Butte

Kwagunt Butte, at right, surmounts a series of Precambrian sedimentary rocks in which some of the earliest forms of life in Grand Canyon have been found. The uplift of the North Rim's Valhalla Plateau is obvious on the left horizon as the eye travels across to the South Rim and the

hydrocarbons, which are formed from the elements hydrogen and carbon, are in a way the parent compounds from which all other organic substances are made. Atoms of oxygen combine with hydrocarbon molecules to form fatty acids, which in turn combine with other radicals to form all known animal and vegetable fats. When the ratio of hydrogen and oxygen atoms present with carbon are in the proportion of two hydrogen to one oxygen, carbohydrates are formed—all the sugars, starches, and cellulose. If an atom of nitrogen bonds with two hydrogen atoms and then links with a hydrocarbon radical, an amino acid is formed. Amino acids bond together to form proteins. One vital group of proteins that facilitate organic reaction are called "enzymes." And so the building blocks pile up.

Painted Desert at the far right. A major feature here is a "monocline," the steep fold interrupting the uniform incline of the rock at center. The panorama was taken from Chuar Butte, which rises sheer above the confluence of the Little Colorado River with the Colorado River. 26. E-5.

The geometry of these building blocks is unbelievably simple, but the complexity of their permutation is vast. How they progressed from their primary state into cellular form and from there to reproduction and mutation is still largely a matter of conjecture. However, the intermediate stage between the two must have begun in the pools that formed on early rock surfaces. The amino acids and the nucleic acid bases were in solution; hydrocarbons and fatty acids formed a scum on the surface. These protobiological materials must have degraded by the continuous effect of solar and terrestrial energy, only to be replaced by fresh organic material in a continuous cycle. It is reasonable to suppose that the element phosphorus was leached from surface rocks by water erosion at this time. It probably reacted first with radiant energy and then with

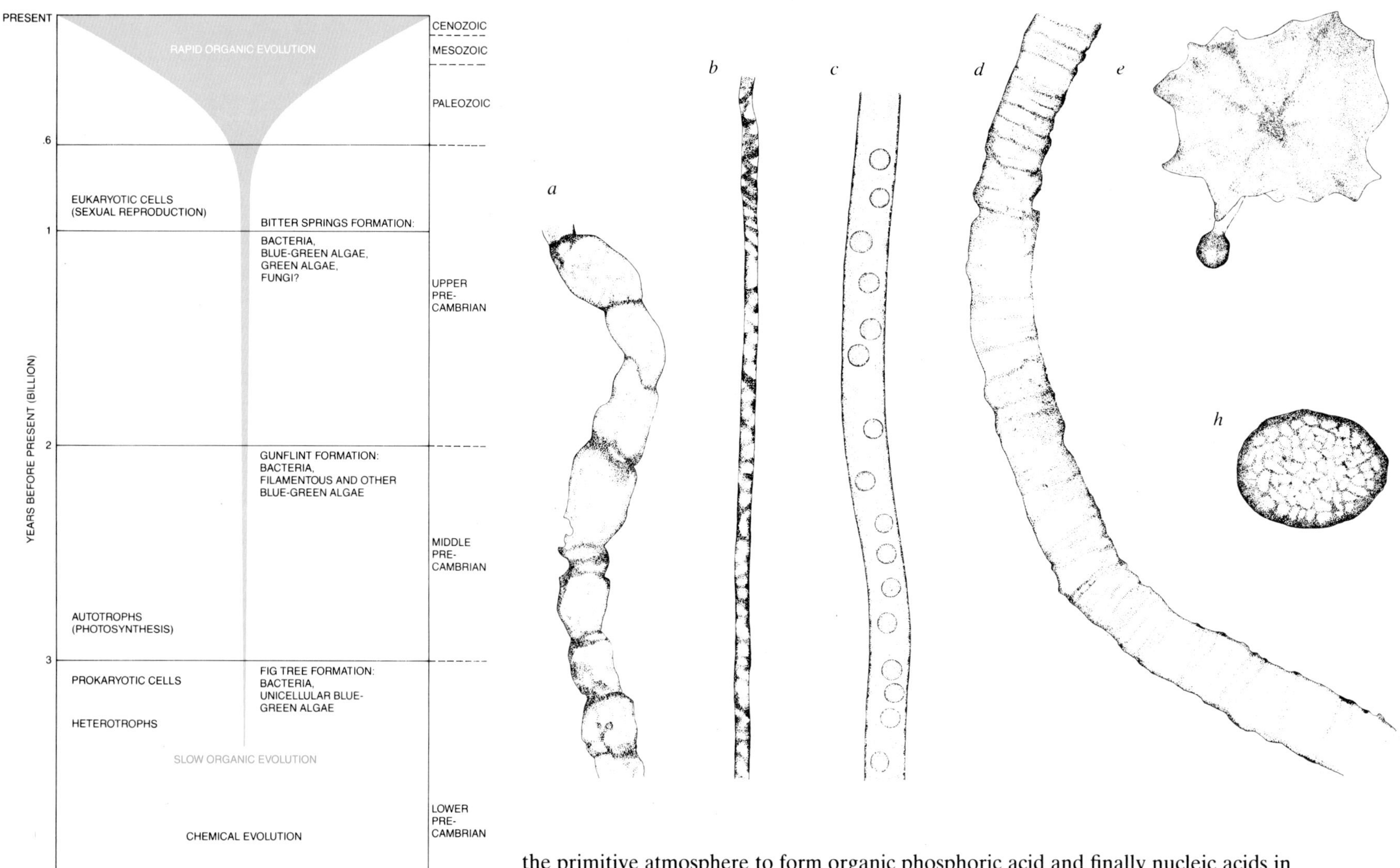

Organic evolution

Organic evolution is presented in terms of successively briefer episodes of biological advance. The Precambrian interval, the earliest and by far the longest, began when the Earth was formed 4.6 billion years ago and ended 600 million years ago with the beginning of the Paleozoic Era. Once organisms with eukaryotic, or truly nucleated, cells evolved, hastening evolutionary progress in late Precambrian times, the number of species multiplied explosively.

the primitive atmosphere to form organic phosphoric acid and finally nucleic acids in combination with the other substances in the pools. These were the base materials that later formed the long helical molecule called DNA, which records and copies all the necessary genetic information for organic reproduction. Fortunately for the process of evolution, the DNA molecular memory sometimes falls short of total recall. This occasional lapse changes the character of the copy it produces, and the copy may prove to be more successful in life than the original. In this case it is the faulty copy, the mutation, which predominates until it too is superseded.

There are a number of ways in which cell-like aggregates of these building blocks could have formed. Oparin suggested that so-called "colloidal suspensions" caused large water-bound molecules to form. The reader will remember that Miller and Urey's primordial soup was cloudy; this was because extremely fine particles of silica from the glass walls of the reaction flask, so fine that they could not sink, had formed a colloidal suspension. Another way of forming a globular mass, suggested by biochemist Sydney Fox, would have been for the desiccated remnants of a primordial pond to have been heated and then slaked with water, for protein substances can be formed if dried amino acids are heated on a chunk of lava and washed off with water. J.D. Bernal proposed that the primordial organic materials reacted with the aluminous clays that

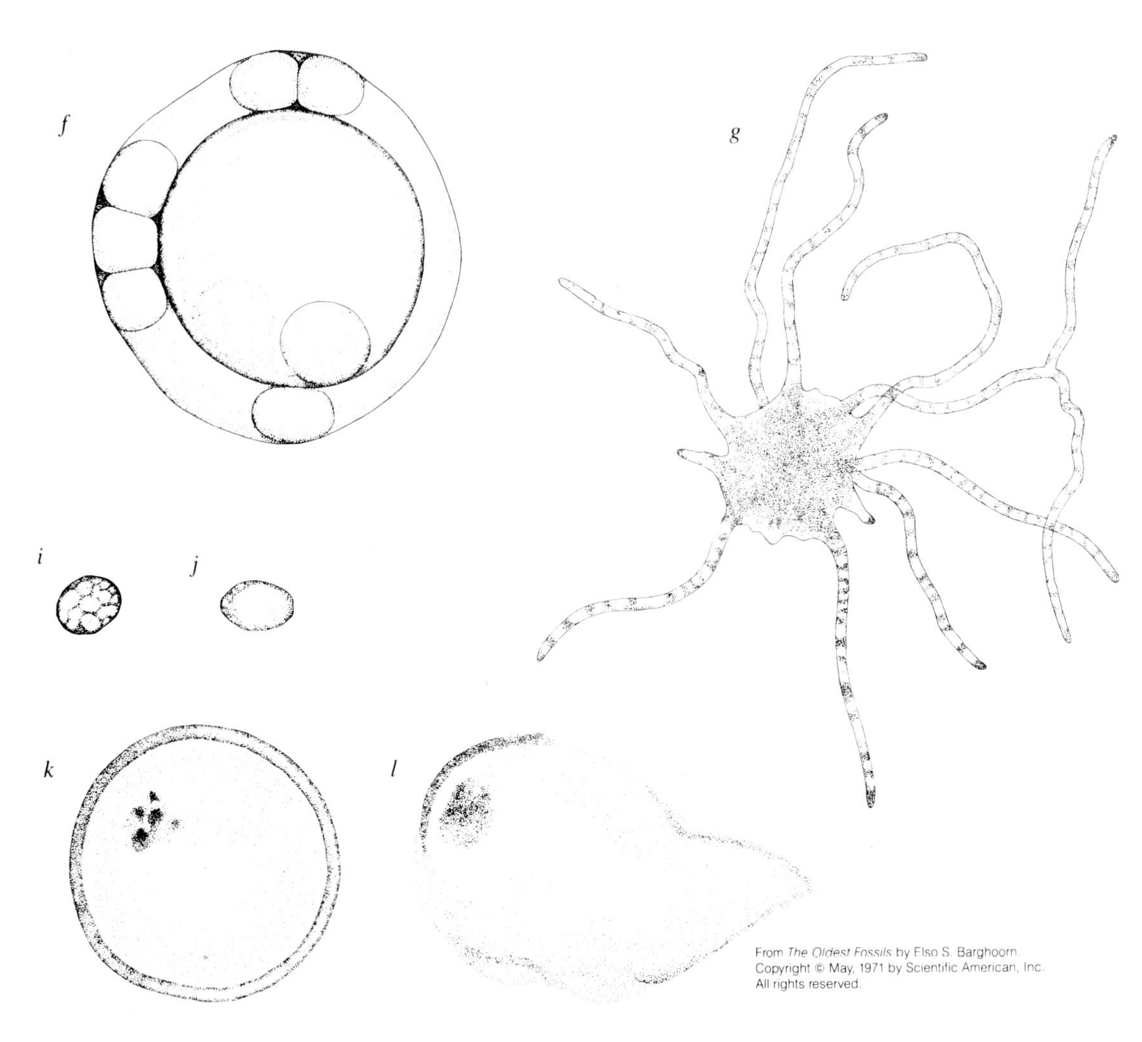

Diversity of form

Everyone has a natural curiosity about the appearance of the earliest known forms of life on Earth which have so far been identified. The drawings of microfossils in this illustration were prepared for Elso S. Barghoorn, a renowned scientist specializing in the identification of microfossil structures. Each reconstruction represents an enlargement 2,500 times that of the original. They are about two billion years old and originated in Precambrian rock on the shores of Gunflint Lake in northeast Minnesota. Illustrations (a) to (c) are structures of filamentous blue-green algae and are believed to be among the earliest organisms capable of photosynthesis. The hydra-like (e) has a known modern counterpart—certain soil organisms, but (f) seems to have failed to survive Precambrian times. Organism (g) is one of two species with an array of filaments which have been found, and (i) and (j) are cellular-like structures which differ chiefly in surface markings. Finally, (k) and (l) are unicellular organisms which might have had nuclei. In fact (l) looks as if it was fossilized during the process of cell division—if so, it suggests that multicellular organisms may have developed earlier in the course of evolution than has hitherto been thought likely.

formed the early sediments, and it is a fact that such clays do react to form biological polymers. And finally, Clair E. Folsome has theorized that the scum of hydrocarbons and fatty acids that formed on the surface of free-standing water clustered into globules with membrane-like surfaces, which enclosed a minute portion of the solution of other organic compounds present in the pool.

All these possibilities seem viable when considered in the light of the geologic processes of primordial erosion described in previous chapters, and when one takes into account the hundreds of millions of years that passed between the precipitation of water, the formation of the oceans, and the appearance of life. Paleontologists have the greatest difficulty in discriminating between organic microstructures preserved in the fossil record of the oldest sedimentary rocks and those of the first living cells. But whatever the explanation for the transition from "organic" to the "living" cell, there is no doubt about the fact that the living cell evolved in two discrete stages, the "prokaryotic" and the "eukaryotic."

Prokaryotic cells first appeared about three billion five hundred million years ago and eukaryotic cells about two billion years later, at about the time that the oldest sedimentary rocks in Grand Canyon were being formed. Prokaryotes were the first organisms on Earth. They were unicellular (single-celled) bacteria and blue-green algae—termed "anaerobic" because they were not dependent on the existence of free oxygen in the atmosphere. Eukaryotes were the first multicellular organisms on Earth. They formed the true algae, many of which are also single-cell organisms, and plants and animals which are multicelled. Eukaryotes are "aerobic"—absolutely dependent on the presence of free oxygen.

Prokaryotic blue-green algae developed when the Earth's atmosphere did not contain any free oxygen. But oxygen was produced by the algae as a by-product of "photosynthesis"—the conversion of water vapor and carbon dioxide into the organic compounds needed for growth. Over hundreds of millions of years the production of oxygen by this means was sufficient to produce free oxygen—oxygen which did not immediately combine with other substances. This process was augmented by the splitting effect that the energy of light has upon molecules of compounds in a gaseous state—a physical phenomena called "photodissociation." The combined effect of both means of oxygen production led to overproduction, in the sense that everything exposed to the atmosphere that could combine with oxygen did so. As the production process continued unabated, it resulted in an excess of oxygen in the atmosphere which was sufficient to allow the formation of a super-oxygen atom called "ozone," produced naturally in the atmosphere by lightning. Together with carbon dioxide and water vapor, ozone accumulated in the stratosphere to form a layer about fifty miles above the Earth's surface. This mantle of gases reduces the permeability of the upper atmosphere to ultraviolet light—a form of radiation that damages living organisms. The overall and very gradual change in environment represented by all these events led to the development and then to the proliferation of eukaryotic cells.

The sterilizing effect of ultraviolet light on the Earth's surface before the advent of the ozone layer makes it improbable that living cells formed in the shallow pools of primordial Earth. It is more likely that life started in the protected environment of the early sediments at the bottom of the developing seas where prokaryotes formed from protocells of organic matter, which were sluiced from igneous rock surfaces by rain.

The oceans as well as the atmosphere were in a state of evolution. Igneous rocks were being leached by weak hydrochloric acid—rain in an atmosphere with a high content of carbon dioxide. The elements silica, sodium, potassium, magnesium, aluminum, and others were released to form salts that were washed into the depressions in the Earth's surface which were to become the oceans. The pattern of inorganic reactions of sea water with sediments introduced new materials. The sediments formed into rocks that became exposed to continental erosion by uplift and were redeposited in the seas. In all, it is believed possible that the whole sedimentary mass has been built up, eroded, and reformed five times during the course of geologic time. Although speculative, it is also believed that the composition of the oceans varied little once the cycle had become established, and hardly at all for the last two billion years. And yet by the present-day processes of erosion, the amount of dissolved silica in the oceans of the world could be doubled in twenty thousand years, and the amount of sodium in seventy million. How is it that the oceans have remained in a steady state of composition for so long? The answer is that the "hydrosphere," as the oceans are collectively called, is not a receptacle that accumulates continental wastes. It is an essential part of the manufacturing mechanism of the Earth, continuously returning its products to the landmasses for reprocessing.

In effect the hydrosphere (the oceans), the lithosphere (the crust), and the atmosphere are intimately linked inorganic vehicles which contribute the materials that promote an organic mass called the "biosphere." All four "spheres" obey the same laws of physics and chemistry and therefore, in time, automatically adjust to changing conditions by reaching a state of dynamic equilibrium. The biosphere is at once the simplest and the most complex component of the quartet. It is the simplest in the sense that 99.99 percent by weight of everything that has ever lived developed from a combination of four abundant and seven common elements; about thirteen trace elements account for the balance of 0.01 percent. None of this material was "created." It all originated from the other spheres, and their total mass was reduced by their contribution to the biosphere's development. The biosphere is also the most complex component of the quartet in the sense that several thousand million living species have evolved since life began. But the evolution of the biosphere as the medium of life on Earth was and still is the product of an environment that resulted from interaction between all four spheres together with the input of solar energy. The spheres are at once inseparable and interdependent. Water and sunlight are their "life's blood."

Water freezes and boils at specific temperatures that depend on the pressure of the atmosphere, which is primarily dependent on the strength of Earth's force of gravity. Gravity determines the amount and therefore the total weight or pressure of the gases retained by Earth to form an atmosphere. Solar energy provides a constant supply of radiant heat and light to both the Earth's surface and atmosphere. In combination, these physical factors determine both the boundaries of life's existence on Earth and its form. Assuming solar energy output to have been reasonably constant since life began, if Earth's gravity had been a relatively few percentage points higher or lower than it actually is, life would have evolved differently. If the Earth's atmospheric pressure had been lower for much of the time, water might mainly have been present as a solid. And if the pressure had been higher, water might have been present mainly as vapor. The world might have been permanently icebound, with life confined to the oceans, or it might have been an extremely hot and humid greenhouse favoring only life-forms that can survive at high temperatures.

The fact is that life developed in forms that can exist at both extremes of temperature, the freezing and boiling points of water. But it only developed explosively when global temperature maintained an average of about fifty degrees Fahrenheit for a prolonged period. This does not suggest that fifty degrees is the temperature at which life evolves. It means that a prolonged period of stable temperature on a worldwide basis allowed life appropriate to that band of temperature to develop.

One supposes that nothing could be more speculative than estimating conditions of climate since life began, but paleoclimatologists make such estimates from a variety of sources and particularly from oxygen isotope radiometry of "cherts"—an extremely fine form of siliceous rock sometimes found in potato-shaped nodules. Some specimens of chert, formed about three thousand million years ago, contain isotopes of oxygen that suggest a climate with a global mean temperature around one hundred and fifty degrees Fahrenheit. During the following thousand million years the average global temperature gradually fell to a mean of about eighty degrees. An average of some fifty degrees was reached about thirteen hundred million years ago, very roughly about the time that the oldest sedimentary rocks in Grand Canyon were formed.

Mean temperatures are misleading as any more than an indication of a general trend in climatic conditions over prolonged periods. Extremely significant details are obscured by vast generalization when using a scale spread over thousands of millions of years. There were, of course, periods of extreme temperature variation from the mean. The first ice ages that are definitely known to have occurred took place during the middle of the Precambrian Era. Dates for rocks with glacial evidence clearly marked upon them cluster around twenty-three hundred million years ago and some rocks suggest the possibility of glaciation several hundred million years before that. Warm climates are evident from the earliest Precambrian sedimentary rock formations and some of the Paleozoic rocks of Grand Canyon, the Mesozoic rocks of Zion, and the Cenozoic of Bryce. Many of these formations clearly demonstrate periods of aridity. But the overall picture of a falling average global temperature that steadied around a mean of fifty degrees about thirteen hundred million years ago is still valid. It indicates no more than a trend and suggests a beginning date for the period of comparative climatic stability that made the rapid evolution of life possible.

This coincided with the three other equally important trends for the prospect of complex life on Earth that have been previously discussed. In the course of two billion years, prokaryotic organisms had photosynthesized enough oxygen to change the composition of the atmosphere, allowing a type of prokaryotic cell to develop that was aerobic. This in turn led to the evolution of the first eukaryotic cells, which are absolutely dependent on oxygen. The whole destiny of life on Earth was determined by this sequence of events, for in the biological sense, sex and competition for existence began with the evolution of the eukaryotic cell.

Eukaryotes are membranous cells that reproduce by division of a nucleus, which contains the essential genetic ingredient DNA. Prokaryotes do not have a nucleus; the cells cannot divide. They replicate by a process called "binary fission"—an expression used to describe the way in which some organic molecules duplicate themselves by splitting into two equal parts. Because of the prokaryotic cell's limitations, for about two thousand million years life-forms had been restricted to unicellular organisms—algae and bacteria. With the evolution of the aerobic eukaryotic cell, complex multicellular organisms formed for the first time. The possible permutations between the genetic information blocks of such eukaryotic cells are so vast as to be considered infinite. The prospects for eukaryotic life-forms were therefore also infinite within the strictures of environment and available food sources. Those forms of eukaryotic cellular organisms that were best at adapting to their living circumstances survived those that could not. Whichever organisms were best at metabolizing available food and, as shortages developed, were best at competing for it, were the progenitors until their progeny were superseded by more successful mutations. The process of natural selection by survival of the fittest to live had begun.

The unicellular algae found in the Bass Limestone below Hance Rapid in Grand Canyon, and discussed at the conclusion of the last chapter, are formed from prokaryotic cells. The earliest multicellular eukaryotic organisms yet positively identified in the rocks of Grand Canyon are from the Chuar Formation in the region of Kwagunt Butte (see illustration page 119). They are microscopic gourd-shaped cells believed to be about eight hundred million years old. Many other specimens of

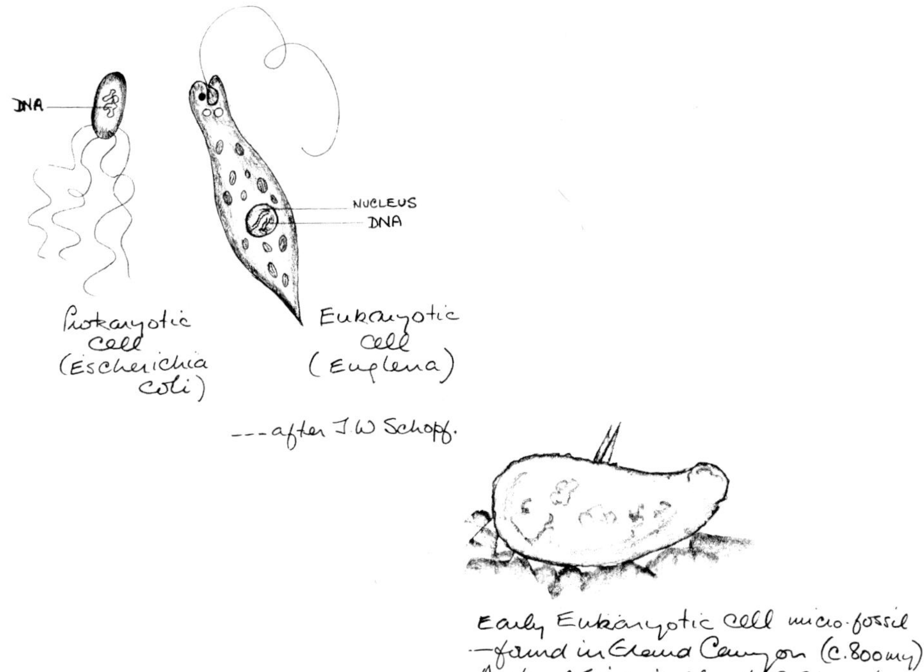

different shapes and sizes have been found in the Precambrian sedimentary rocks of the Canyon, which some geologists have interpreted to be organic, but which others do not accept. But further links in the evolutionary chain from fifteen hundred million years ago to the beginning of the Cambrian and the appearance of the trilobites have yet to be positively identified there. When and if they are eventually found, they will almost certainly take the same or similar form as Precambrian fossils discovered elsewhere.

All the very early multicellular organisms appear to have been soft bodied and all lived in the sea. Some crawled, some floated, some swam, and some burrowed. They were jellyfish and worm-like creatures. One Precambrian fossil find in Australia proved to be a small annelid worm which had a flexible body nearly two inches long with a horseshoe-shaped head and about forty pairs of lateral body projections ending in needle-like spines. It is a distinct possibility that this creature is related to the trilobites.

How prokaryotic cells evolved into eukaryotes, and eukaryotic cells into multicellular organisms such as the annelids, is a matter of profound debate. "Symbiosis" or "mutualism"—the living together of two dissimilar organisms to their mutual advantage—offers both a clue and one of many possible explanations. Because the nucleus of eukaryotic cells contains several parts, biologists suggest that these might at one time have existed as separate, unlike, prokaryotic cells that assumed a symbiotic relationship to fulfill a mutual need. The union could have been between bacteria dependent on oxygen and algae capable of photosynthesis. The association may have been so successful that it was perpetuated in the form of a proto-eukaryotic cell. If one now makes the assumption that this new class of cell assumed further symbiotic relationships until it achieved amoeba-like characteristics, the next

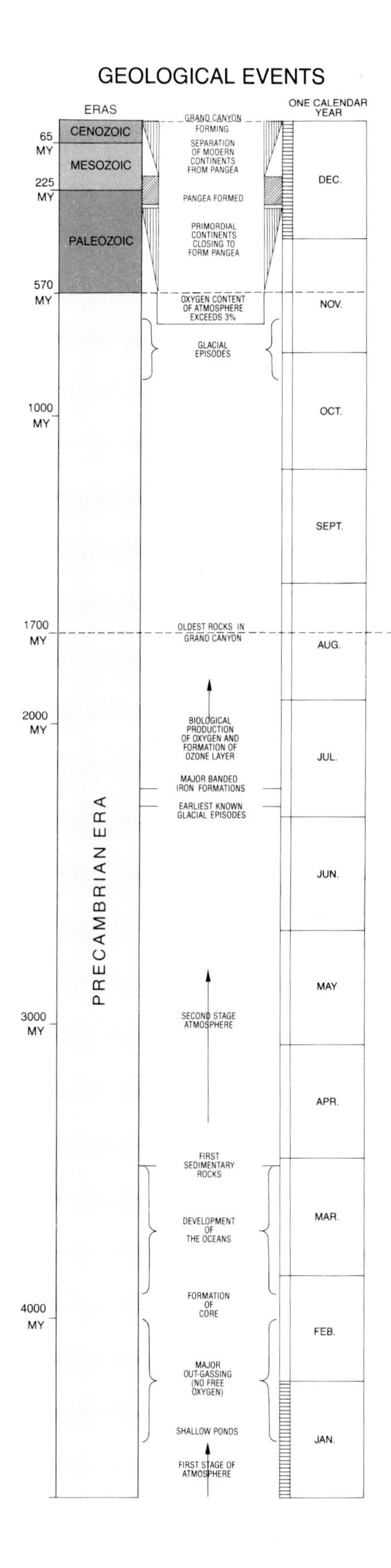
GEOLOGICAL EVENTS
ERAS
ONE CALENDAR YEAR
GRAND CANYON FORMING
CENOZOIC
65 MY
SEPARATION OF MODERN CONTINENTS FROM PANGEA
MESOZOIC
DEC.
225 MY
PANGEA FORMED
PALEOZOIC
PRIMORDIAL CONTINENTS CLOSING TO FORM PANGEA
570 MY
OXYGEN CONTENT OF ATMOSPHERE EXCEEDS 3%
NOV.
GLACIAL EPISODES
1000 MY
OCT.
SEPT.
1700 MY
OLDEST ROCKS IN GRAND CANYON
AUG.
2000 MY
BIOLOGICAL PRODUCTION OF OXYGEN AND FORMATION OF OZONE LAYER
JUL.
MAJOR BANDED IRON FORMATIONS
EARLIEST KNOWN GLACIAL EPISODES
PRECAMBRIAN ERA
JUN.
MAY
3000 MY
SECOND STAGE ATMOSPHERE
APR.
FIRST SEDIMENTARY ROCKS
DEVELOPMENT OF THE OCEANS
MAR.
FORMATION OF CORE
4000 MY
FEB.
MAJOR OUT-GASSING (NO FREE OXYGEN)
SHALLOW PONDS
JAN.
FIRST STAGE OF ATMOSPHERE

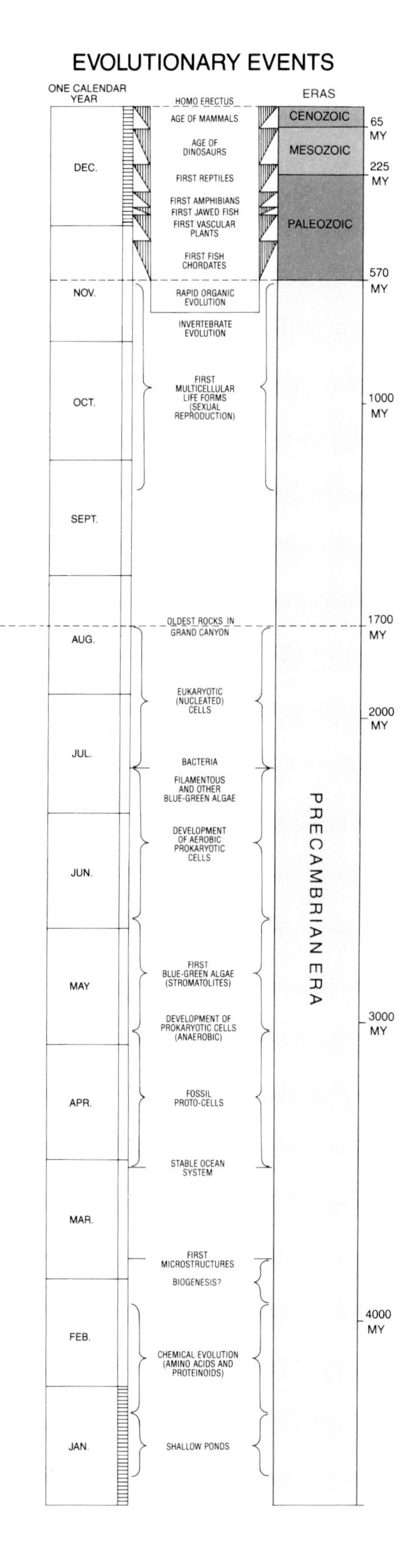
EVOLUTIONARY EVENTS
ONE CALENDAR YEAR
ERAS
HOMO ERECTUS
AGE OF MAMMALS
CENOZOIC
65 MY
DEC.
AGE OF DINOSAURS
MESOZOIC
225 MY
FIRST REPTILES
FIRST AMPHIBIANS
FIRST JAWED FISH
FIRST VASCULAR PLANTS
PALEOZOIC
FIRST FISH CHORDATES
570 MY
NOV.
RAPID ORGANIC EVOLUTION
INVERTEBRATE EVOLUTION
OCT.
FIRST MULTICELLULAR LIFE FORMS (SEXUAL REPRODUCTION)
1000 MY
SEPT.
OLDEST ROCKS IN GRAND CANYON
1700 MY
AUG.
EUKARYOTIC (NUCLEATED) CELLS
2000 MY
JUL.
BACTERIA
FILAMENTOUS AND OTHER BLUE-GREEN ALGAE
PRECAMBRIAN ERA
DEVELOPMENT OF AEROBIC PROKARYOTIC CELLS
JUN.
FIRST BLUE-GREEN ALGAE (STROMATOLITES)
MAY
DEVELOPMENT OF PROKARYOTIC CELLS (ANAEROBIC)
3000 MY
APR.
FOSSIL PROTO-CELLS
STABLE OCEAN SYSTEM
MAR.
FIRST MICROSTRUCTURES
BIOGENESIS?
4000 MY
FEB.
CHEMICAL EVOLUTION (AMINO ACIDS AND PROTEINOIDS)
JAN.
SHALLOW PONDS

The young finger canyons of Kolob, Zion

This is a part of Zion which too few people venture to see. Located on the northwestern boundary of the National Park, it is a rewarding visit. It is bewildering to realize that these young finger canyons, the rocks from which they evolved, and the formations that extend several miles beneath them, just did not exist when the first life forms were evolving on Earth. It will be surprising to many that such an arid country can be deep in snow. In winter this part of Zion is actually closed to the public, and the author had to fly in to this location by helicopter to take this panorama on a glorious, clear, midwinter day.
27. D-3.

step might well have been cell division as a better means of reproduction than replication by binary fission.

There are a vast number of present-day mutualistic relationships in nature that lend support to the first link in this possible chain of events. One of the most interesting examples of present-day symbiosis in the context of evolution is that some plants (eukaryotic cells) contain a type of bacteria (prokaryotic cells) in their seeds that appear to maintain a separate identity and yet are inherited by the plant's offspring in some way not yet understood. Cell division, the second link in the chain, whenever and however it occurred, was one of the most important of all evolutionary events, for it led directly to the development of eggs and sperm.

At some distant time during the early dawn of life on land, an anaerobic single-celled fungus that required organic food formed a symbiotic relationship with an aerobic algae that required inorganic nutrients. Although other inventions of nature overtook this life-form in complexity and in almost every other respect, it is a fact that this particular relationship which we call "lichen" is possibly one of the oldest forms of land life. It exists high on the highest mountains. It is found in the Antarctic. It tenuously exists in the Sahara. And if that exobiologist from another planet in our galaxy did indeed land a version of *Viking Lander* on Escalante Butte in Grand Canyon, the first shovelful of surface material collected by the vehicle would almost certainly prove life's existence on Earth by the presence of lichen.

Lichen — a symbiotic relationship between algae & fungi

VIII
THE MAGIC NUMBER

The Araucarioxylon

In the Painted Desert there is a petrified forest of hundreds of trees. This fossil tree is one of the finest specimens. It lived nearly 200 million years ago in Late Triassic times. The tree was possibly several hundred feet tall, with a diameter of some ten feet at its base. Modern descendants of the Araucarioxylon are Monkey-Puzzle trees, which grow naturally only in the Southern Hemisphere.
28. G-7.

A majestic ocean roller raced shoreward. It curled, crested, and hovered all shimmering green and translucent in the bright light of a noonday sun. Driven by wind and tide, it tripped over a shallow reef and broke into pounding surf. Momentum finally spent, it flung a shallow foam-flecked wave to lick the naked beach. Its roar had not been heard. Its regal progress had not been seen. It received no greeting but the silence of the glistening, sunlit sand. In those Precambrian times, life on land had not begun.

Winds, tides, waves; days and nights, months and years; summer, autumn, winter, spring, or just seasons are the pulses of Earth to which terrestrial life has had to conform. In response, all organisms have developed corresponding rhythms. There are "sleeping" activities and "waking" activities tied in to the nocturnal or diurnal modes of the organism. There are monthly, biannual, and annual reproductive cycles with gestation periods that correspond to produce the succeeding generation at the time that best ensures both its survival and the continuity of the species. Indeed, every living system is so responsive to Earth's (and the Moon's) physical rhythms that one might combine cause and effect to conclude that each system incorporates a biological clock.

Part of the spray from the spuming wave evaporated. Some of those tiny droplets may have contained unicellular fungi and particles of green algae which had already formed a symbiotic partnership near the surface of the sea. One day, on some wave, perhaps the first of countless of trillions of pairs survived. Maybe it was on the surface of a sheltered spray-damp rock high up on a cliff, or maybe it was far inland. But survive it did, in just the right ecological niche. The first lichen had landed. The magic number which perpetuated its kind was "two"—a number which was to become even more significant. The wave licked the naked beach and deposited the trillionth diatom on the glistening sand. But there was a difference. This particular unicellular algae could insinuate itself below the surface of the sand until the tide receded and then emerge from the damp sand to photosynthesize in the sun. Perhaps this was the first faltering step the sea plants took in their transition to the land plants.

With twenty-five hundred million years of primary evolution behind them, eukaryotic life forms developed at an explosive rate at the end of the Precambrian Era and the beginning of the Cambrian Period. In time, life's diversity was only rivaled by its complexity. Such life forms are categorized into "kingdoms," plants and animals constituting the more familiar ones. Each is subdivided into "phyla," and phyla into "classes," and so on down the line of "orders," "families," "genera," and "species." The history of animals is known best because of the extensive fossil record left behind. That of fungi is the least well known because comparatively few fossil traces have been found.

The development of life forms was certainly affected by the assembly and the breakup of the primordial continent of Pangea. At different times some life forms were almost annihilated by changes in the environment, which may well have been related to changes in their degree of latitude. The illustration on page 23 suggests the relative shape and relationship of the major continents and oceans from Cambrian times to the present day. First there was an assembly of pre-Pangean continental masses. Then the amalgamation of these continents into one—Pangea. Finally Pangea divided into Gondwanaland and Laurasia and then into the smaller continental masses with which we are familiar.

As the continents drifted, collided, separated, and drifted again, the configuration of ocean currents changed dramatically. This altered the pattern of food distribution in the seas and caused substantial fluctuations in climate. It is believed that in early Cambrian times all life was aquatic and existed mainly in shallow marine environments. The reefs near tropical coastlines were the scene of lively activity and were therefore the most susceptible to environmental change. They were probably the first interdependent animal and plant communities on Earth. Such communities, with their associated physical environments, are called "ecosystems."

The margins of a reef are seldom more than two hundred feet below the surface of the sea and are built of sand stabilized by plants and animals anchored to its surface. The temperature of the clear and richly oxygenated water remains in the eighties Fahrenheit throughout the year (well above the global average), and varies by approximately five degrees Fahrenheit through winter and summer. The primitive algae-type seaweed called "stromatolites" formed the early reefs, and some of these are evident in the Precambrian rocks of the Grand Canyon. The first crude ecosystems formed in early Cambrian times when stony, sponge-like animals and the first trilobites lived in the stromatolite reefs. But it wasn't until the middle of the Ordovician Period (which has no representative rocks in Grand Canyon), about one hundred million years later, that reef communities really began vigorously to develop with the evolution of the animals with exoskeletons, which are called "corals."

The body of a coral animal consists of a soft, hollow, cylindrical structure attached at its lower end to the surface of a reef. The tentacles at the top can usually extend to gather food. The animal is carnivorous and has developed stinging cells in the region of its tentacles that paralyze its prey. Corals grow a calcium carbonate skeleton by ingesting seawater and extracting the compound. When they die, their skeletons are left attached to the reef. In the Paleozoic Era the corals, with millions of others of their kind, and with the vast number of seashells left by the brachiopods and marine mollusks, were primarily responsible for limestone formations. Today there are about twenty-three hundred species of stony, horny, and thorny types of coral. But it is the stony species that originated in Paleozoic times that holds a special interest, for some of these have proved to be paleontological clocks.

When living, the upper and soft part of the animal, called a polyp, resembled a small sea anemone. It exuded a circular-shaped deposit of calcium carbonate from its base onto the rock surface. Such deposits accumulated in a cylindrical shape, and as the polyp grew larger, the top "rings" of the skeleton increased in size so that its overall shape became conical. In 1963, John W. Wells of Cornell University suggested that the fine rings within the annual bands of growth he had observed in the structure of such Devonian corals in fact recorded the daily growth rate of the animal. Wells counted an average of four hundred daily rings within annual bands, which implied that with four hundred days in the Devonian year, the Devonian day must have been two hours shorter than present days.

The length of the day depends on the speed of rotation of the Earth on its axis. At the time of Wells' discovery, it was known that the speed of Earth's rotation is gradually slowing down. The cause of this is attributed to the drag effect of Earth's ocean tides, which are raised by the Moon. Calculations based on the effect of tidal friction on the speed of rotation had suggested a gradient of shortening days going back to primordial times, but until the time of Wells' discovery there had been no way of independently checking these estimates. Subsequent work with corals of Paleozoic, Mesozoic, and modern origin produced results that matched the tidal friction calculations. In round figures, the day has been increasing in length by about half an hour per one hundred million years. This means that in Cambrian times there were about four hundred and twenty-five days in each year, and that each Cambrian day lasted about twenty hours and thirty-six minutes. On the same basis, the length of a Precambrian day of approximately twelve hundred million years ago was about eighteen hours, with four hundred and eighty days in one year. These conclusions have a twofold significance for the paleontologist. They add to the general picture of ancient environments—shorter and more frequent days and nights, perhaps stronger tidal movement—and they suggest that other living organisms of the past such as the stromatolites may have had similar characteristics—they too may have been paleontological clocks.

The ocean tides and waves may also have played an important part in evolution in the development of the invertebrate marine animals into those that became invertebrate land animals. Whichever reef creatures from the phylum *Arthropoda* (from which trilobites are the oldest-known representatives) survived the alternate conditions of immersion in seawater and exposure to air and direct sunlight lived to produce the next generation of their kind. The order in which arthropods are classified suggests how they might have evolved. First the crustaceans: fairy shrimp, barnacles, crabs, lobsters, and crayfish. Then the arachnids: scorpions, spiders, mites, and ticks.

Then the myriapods: millipedes and centipedes. And finally the hexapods: the vast number of species of insects ranging from dragonflies to butterflies and moths, and from termites and lice to mosquitoes and fleas.

From Cambrian times to the present, reefs have suffered four long periods of decline followed by resurgence and reestablishment. These periods have always been accompanied by the appearance of new species of fossilized animals in the limestone formations for which their shells or skeletons were primarily responsible. Corresponding intervals of decline or even hiatus are not only apparent at Grand Canyon, Bryce, and Zion but throughout Earth. In Grand, Zion, and Bryce canyons times of decimation coincide with major unconformities because at these times the rocks were raised above the surface of the sea. The unconformities referred to are those between Muav Limestone and Temple Butte Limestone, and the Temple Butte and Redwall Limestone in Grand Canyon; between the Kaibab Limestone and the Moenkopi Formation at Zion; and the Kaiparowits and the Wasatch Formations at Bryce.

Each of these four periods appear to coincide with major changes in the configuration of the continents. The first two unconformities occurred as the land masses that had separated out from the Precambrian supercontinent were moving together to form Pangea. The third occurred after Pangea had formed and had begun to separate into Gondwanaland and Laurasia. And the last came about when the present continents had taken shape and had begun drifting into their present positions. These movements altered coastlines, formed mountains, changed ocean currents, and caused

Aspen at Cedar Breaks

Cedar Breaks is an area between Bryce Canyon and Zion Canyon which is locally renowned for the wonderful autumn colors of its aspen groves. The aspen turn color for a few weeks each year during late September or early October. But at their elevation of 8,000 to 9,000 feet the pageant of trembling, brightly colored leaves can be abruptly ended by a high wind or by one brief fall of early snow.
29. C-4.

ice caps to form on landmasses in the south polar region and in the sea at the north polar region. The planet's climate was affected. Continental warping and ocean bed deformation caused shallow seas to be formed and drained again. Also, ice at the poles and on the continents lowered sea level as it formed, and raised sea level when it melted.

Reefs are essentially shallow marine communities and their ecology is very sensitive to environmental change of any kind, and particularly those of depth and temperature. It therefore seems probable that continental movement was a prime cause of the environmental changes that are reflected by the four protracted periods of reef community decline and resurgence. Among the marine plants, the stromatolites survived the first two and then languished during the third at the end of the Permian Period. Eukaryotic green algae were prolific from about three hundred million years ago but also waned at the end of the Permian. Coral-like algae that secreted calcium carbonate developed between the first and second episodes and have survived down to the present time. The animals associated with stromatolites became more diverse between each episode, but few species have survived in their original form.

At some time in the past, marine plants that grew so prolifically in reefs near seashores evolved into land plants. The evolutionary problem to be solved in every transition of life from sea to land was how to survive out of water. Tidal movement must have frequently exposed part of some reefs. One can reasonably assume that as a consequence, plants that evolved on such reefs were able to survive in either seawater or in the atmosphere for part of their lives. The key to survival seemed to be that some part of the lower plant remained moist while the rest was exposed to sunlight and that there should be moisture-conductive tissue between the two extremities.

At the turn of the century a French botanist, Octave Lignier, put forward the theory that a primitive green alga called "chlorophyte" was the first to master this evolutionary step. Chlorophyte is so named because it contains chlorophyll, the essential ingredient that enables plants to photosynthesize carbohydrates from sunlight. Lignier suggested that the green alga had evolved a primitive root system that had secured its survival when it was left high and dry, first by receding tides, and later by permanently receding seas. He went on to suggest that a further refinement caused the plant to lift its reproductive parts into the air. This step, he suggested, was succeeded by the development of bundles of ducted, or "vascular," tissue that aided circulation of water from the roots to the head of the plant. The further development of a stem encouraged the evolution of spatulate, or leaf-like, growth at the extremities so that photosynthesis by sunlight increased to cope with the alga's needs. Lastly, a cuticle formed over the leaves to protect the whole system from excessive evaporation.

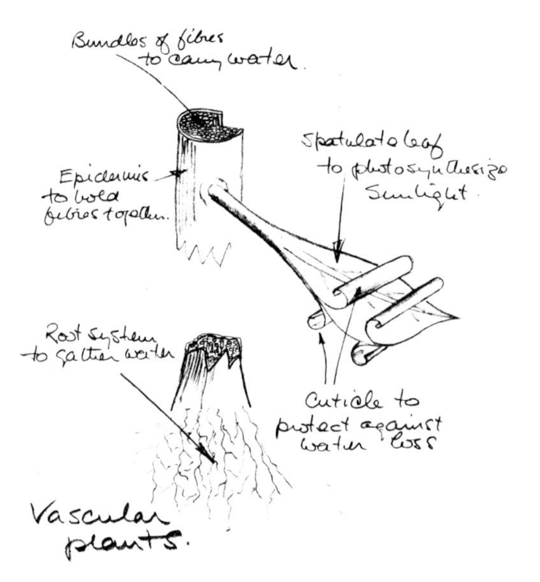

At one time Lignier's theory was not generally accepted, but today it is broadly held to describe the most likely sequence of events that led to the evolution of vascular plants. Lignier's ideas seemed to be confirmed by the discovery of the four-hundred-million-year-old Silurian Period fossil *Cooksonia*, a primitive *Psilopsid* which is the oldest known rooting plant, and by the discovery of the fossils of two other primitive plants in Devonian rocks. These plants had branched, spine-like leaves with reproductive spore sacs at their ends. *Psilophyton* and *Rhynia*, as they were named, had a number of characteristics that correspond to the principal features of present-day club mosses, horsetails, and ferns—all of which are spore-reproducing plants, as well as to the earliest seed-bearing plants. Club mosses and horsetails did not progress far along the evolutionary path. Ferns were more successful, but they too preferred humid conditions. During times of semiarid or arid conditions, and by all accounts these were frequent, they were overtaken by the seed-bearing plants. These had the advantage of being able to adapt to changing environments and led to the evolution of conifers and, much later, to the angiosperms—the flower-bearing plants. There are now more than a quarter of a million species of angiosperms on Earth. They dominate the plant kingdom numerically. All deciduous trees and many evergreens belong to this phylum, because just like the plants we call "flowers," they too are flowering plants with extra large stems, and they reproduce sexually: Pollen grains are produced by male organs, and egg cells form in female reproductive organs.

Among the best known of all fossil plants in the region of our canyon trinity are the trees of the Petrified Forest in the Painted Desert. They were members of thc conifer family and most of them are of the genus *Araucarioxylon* (see panorama pages 132–133). They grew in this region about one hundred and ninety-five million years ago in Late Triassic times, in the Mesozoic Era. Many thousands of petrified logs are found in the Painted Desert, which covers an area about one hundred and fifty miles long and up to fifty miles wide. But the greatest concentration of logs is to be found in the Petrified Forest National Park near Holbrook in northern Arizona. Here they lie in large numbers, haphazardly strewn around in a primordial-looking landscape.

By Late Triassic times the Tethys Sea, so named after the wife of the Greek god Oceanus, had separated the continent of Pangea into the northern subcontinent of

Laurasia and the southern subcontinent of Gondwana (see illustrations page 23). Laurasia later became North America, Greenland, Europe, and Asia north of the Alps and the Himalayas. Gondwanaland became South America, the rest of Asia, Africa, Australia, and Antarctica. Because of this initial division, all living things were very broadly separated into those that eventually evolved in the Northern Hemisphere and those which usually relate to the Southern Hemisphere. The *Araucarioxylon* of the Petrified Forest grew on Laurasia, but the group evidently did not survive the changed environment to which the subcontinent was exposed. The modern descendants of *Araucarioxylon*, the genus *Araucaria*, grow in Australia, the Southern Pacific, and in South America but not (unless artificially planted) in the Northern Hemisphere from which their ancestors were isolated nearly two hundred million years ago.

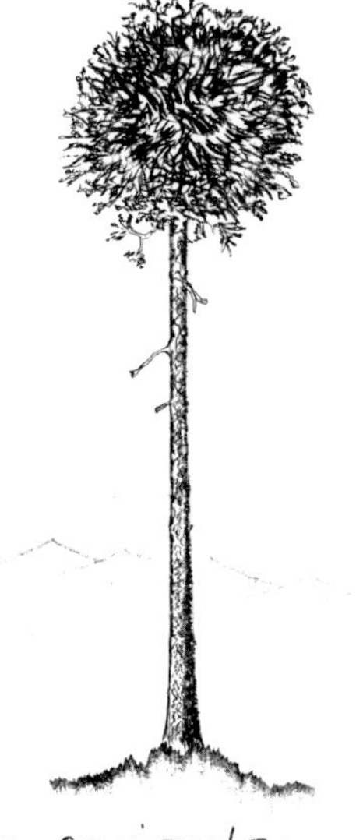

Araucarioxylon — was the trees of the Petrified Forest

The *Araucarioxylon* of the Painted Desert grew in large stands on scattered knolls in a Laurasian landscape that bore a resemblance to the modern subtropical Everglades of Florida. Some *Araucarioxylon* grew to a height of several hundred feet or more with diameters up to ten feet at the base. They lost their crowns and their roots after death, either by rotting or during the process of being tumbled and jostled as they were moved by the muddy floodwaters in which they eventually petrified. The bright colors of the petrified wood are best appreciated when a section of it is polished. The colors are the result of iron oxide, manganese, and occasionally copper and chromium in the siliceous mud in which it was petrified. Many trees broke into sections after they had been fossilized, and this may be attributable to the frequent earthquake tremors that occurred during the Plateau's period of greatest uplift many millions of years after the Petrified Forest had formed. In addition to the trees, there was an abundance of other Laurasian plant life in the area, much of which was also fossilized. The ferns are strongly represented, as are some of the early seed plants, of which the now extinct *Bennettites* are a good example.

The Late Triassic sediments that accumulated in the area of what is now the Painted Desert constituted the Chinle Formation. The Chinle is also extensively exposed at Zion (see panorama pages 30–31) as well as in the region of the Painted Desert. It is a particularly rich source of fossils. This is not because other formations are necessarily less productive, but because of the nature of the Chinle itself. Unlike most other formations in the Plateau region, it is extremely soft, particularly after rain or snow, and is more like hard-packed gray-and-colored talc than rock. It was formed from extremely fine materials—muds, silts, and bentonite (volcanic ash turned to clay)—that have remained soft in spite of the fact that they once were covered by about three thousand feet of other sedimentary rock that has been eroded away. The rate of Chinle erosion is very high, and as a consequence, fossils are frequently washed out of the sediments. These fossils are not just of plants but also of animals. The animals were fish, amphibians, and reptiles, and they were vertebrates—each had a skeleton with a backbone, perhaps the most marvelous of all natural inventions for which evolution was responsible.

The early vertebrates are believed to have evolved from small, streamlined marine animals called "protochordates." These creatures were little more than segmented tubes with a gill-like apparatus at one end and a primitive axial skeleton running the length of the body. They possessed a closed blood-circulation system, and they reproduced their progeny sexually in the form of larvae. The protochordates' closest invertebrate relatives are the echinoderms—the crinoids, starfish, and sea cucumbers.

By the Late Ordovician Period the primitive chordates had developed into armored jawless fish that fed themselves by filtering nutrients suspended in seawater through their mouths and out of their gills. The lampreys are probably the modern descendants of these early vertebrates. The process of natural selection naturally favored those fish that developed hinged jaws, because this broadened the variety of food that an animal could ingest, and therefore improved its chances of survival. The jawed fish that resulted fall into two categories, the rayfins and the lobefins. The former led to the evolution of sharks, skates, rays, and the multitude of bony fish that first appeared in very large numbers in the Devonian Period. Some of these families survived to the present in modified form, and the sharks and rays of today are their descendants. Some did not survive far beyond the Devonian. The fossil fish *Bothriolepsis* found in Grand Canyon is one of these. This was an armored freshwater bottom feeder and one of the first animals to develop lungs. The tropical rivers it inhabited were either tidal, prone to drying out, or both. These environmental factors encouraged the evolution of species that could survive out of water. The genus to which *Bothriolepsis* belonged became extinct and what part such species played in the evolution of amphibia is obscure. But the genus did show that fish were capable of developing lungs. Because lobefins had primitive appendages that could have become limbs, and had the potential for lung development, these and other factors have led to the conclusion that the first amphibians that appeared in Late Devonian times evolved from the lobefins. The earliest amphibian trace fossils to appear in the rocks of Grand Canyon are found in the three-hundred-million-year-old Pennsylvanian rocks of the Supai Group. The tracks that appear on some Supai rock surfaces look as if they were freshly made by a small, short-legged, low-slung animal.

Upper Sonoran spring, Toroweap

The elevation here is about 5,000 feet, and the late May days are pleasantly hot in daytime and cool or even cold at night. But in summer it is often over 100 degrees Fahrenheit during the day. This country may look stark, but, in fact, insects are busily pollinating, and fairy shrimp are mating in the rain pools in the shallow depressions of the Supai Formation rock. Fairy shrimp are an example of how animals can adapt to the harsh climate of the region. The eggs that are fertilized in the spring remain dormant until the fall rains. These eggs then hatch into females, which after several stages of growth lay unfertilized eggs. Some of these may develop into male fairy shrimp that fertilize remaining eggs, which then lie dormant until the next hatching. 30. E-3.

The force that encouraged the lobefins to evolve into amphibians was possibly initiated by the ebb and flow of tides and the subsequent evolution of those animals that survived regular exposure. It could equally well have derived from an instinct to survive when aquatic animals found their watery abodes drying up. This may have provided an incentive for them to try to transfer from a shallow pool to a deeper one, or from a dwindling stream to one that was running high. There could have been other factors that worked in combination; gradual regression of the sea is one possibility, and the availability of food on dry land in the form of plant life is certainly another. Whatever the cause of their dual role, the early amphibians, much like their present-day descendants, the salamanders and frogs, were tied to water for the purpose of reproduction. This seriously limited their evolution.

Reptiles evolved from one of the early amphibians. The prime difference between the two was that the reptiles could lay eggs with leathery casings, and the amphibians could not. No doubt there were countless occasions when amphibian eggs were spawned in water that dried up, or were left high and dry on a beach by a receding wave. This must have occurred over a sufficiently prolonged period, in order for a series of mutations to have enabled a clutch of fertilized eggs with a tough enough membrane to develop and survive to maturity partially or wholly out of water. This rubbery membrane of reptile eggs, which eventually led to the development of the hard

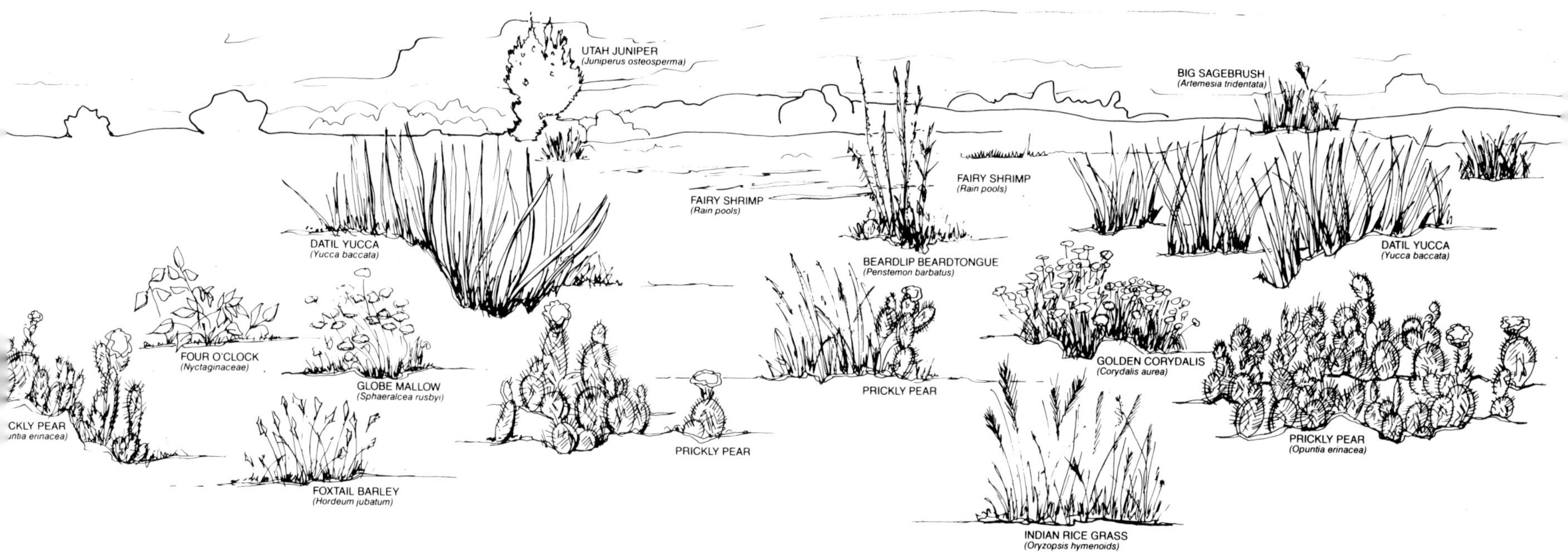
UTAH JUNIPER
(Juniperus osteosperma)
BIG SAGEBRUSH
(Artemesia tridentata)
FAIRY SHRIMP
(Rain pools)
FAIRY SHRIMP
(Rain pools)
DATIL YUCCA
(Yucca baccata)
BEARDLIP BEARDTONGUE
(Penstemon barbatus)
DATIL YUCCA
(Yucca baccata)
FOUR O'CLOCK
(Nyctaginaceae)
GLOBE MALLOW
(Sphaeralcea rusbyi)
GOLDEN CORYDALIS
(Corydalis aurea)
PRICKLY PEAR
CKLY PEAR
intia erinacea)
PRICKLY PEAR
PRICKLY PEAR
(Opuntia erinacea)
FOXTAIL BARLEY
(Hordeum jubatum)
INDIAN RICE GRASS
(Oryzopsis hymenoids)

eggshell of birds, enabled an embryo to develop in an independent, well-protected, and watery world of its own. The embryo was enclosed in a membrane. It had a food supply—the yolk—and a means of disposing of nitrogenous waste—the cellantois. The whole system was enveloped in a protective sac—the chorion, which was the precursor of the placenta.

Evolution of the "shell" egg was followed by rapid diversification of the reptiles. They became primarily land based and developed hard, dry, outer skin. They continued to live in the same estuarine environment as their moist-skinned amphibious relatives. The reptiles and the amphibians competed for food. The reptiles evolved more rapidly than the amphibians as they too became predators as well as competitors. By their superiority in numbers the reptiles forced the amphibians into decline. The animal fossils that are most commonly found in the Chinle Formation of Triassic times are typical of this stage of evolution. The amphibians are most frequently represented by the bones of a salamander-like animal, and the reptiles by a long-snouted beast of crocodilian appearance. Some reptiles developed into ungainly looking animals called "therapsids." The therapsids included subspecies that possessed mammal-like teeth. From this and other evidence it has been concluded that mammals are descended from therapsids. Meanwhile, another reptilian line called the "thecodonts" was developing. The Mesozoic Era was a time during which size and brute force, or guile and speed, became the determining factors of survival. The thecodonts evolved into that legendary lineage of tale and fable called "dinosaurs."

Dinosaurs were the reptiles that dominated the animal kingdom on Earth's surface for one hundred and fifty million years, from Triassic and Jurassic times until toward the end of the Cretaceous Period. Because they were around for such a prolonged period they left many signs of their presence. Their fossilized bones are an occasional and exciting find. But when one comes across three-toed footsteps left by some dinosaur between strides in the sands or muds of its day now indelibly imprinted in rock surfaces, one senses a vivid reality that one never gets from finding an odd fossil bone in the desert or from seeing a reconstructed skeleton in a museum. These trace fossils are to be found in some of the Late Triassic rocks of the Painted Desert and the Jurassic rocks of Zion. During Cretaceous times much of the area was under the sea, so the local record of the dinosaur dynasty faded from the rocks that formed in this locality at that time.

Toward the end of the Cretaceous, perhaps during the course of relatively few millions of years and certainly in evolutionary terms a short period of time, most but not quite all the dinosaurs became extinct. In addition to a sharp division between the herbivores and the carnivores, and differences in the locomotion styles of quadrupeds and bipeds, there was also another natural anatomical division in the types of dinosaur: the reptile-like "saurischia" and the bird-like "ornithischia." Some survivors related to the saurischia evolved into the present-day crocodilians, lizards, and snakes. Some survivors related to the ornithischia are believed to have been warm-blooded bipeds. Their highly evolved descendants of today are also endothermic. They have scaly legs with claws and a wishbone that bear testimony to features that they inherited from their Jurassic antecedent, the "archaeopteryx," which was the animal generally credited with being the very first bird.

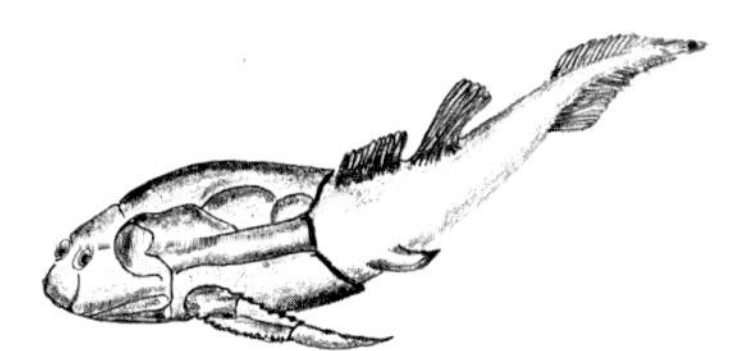

Bothriolepsis
— Devonian freshwater
lobefin.

The dinosaurs have a special fascination for us all. Presumably this is because of their size, although most were probably quite small, their sometimes fierce inclinations, and their diversity. Something new is always being discovered about them. Perhaps this is why the fact of their "sudden" extinction has tended to obscure the other and equally dramatic mass extinctions of animals that are evident in the fossil record. One has to remember that in geological terms the word "sudden" can be applied to a change that has taken place over millions of years. A glance at the stratigraphic chart on page 66 is sufficient to identify quickly a number of major unconformities between the rocks of our canyon trinity in the last six hundred million years. Several of these coincided with periods of major mass extinctions all over the Earth, of which those that occurred at the end of the Devonian and Permian Periods are an example.

The forces that put an end to the dinosaurs at the end of the Cretaceous Period not only devastated other animal families at the same time, but were perhaps of a recurrent character. When paleontologists analyzed the ebb and flow of life's resurgent development since Cambrian times, they concluded that the destiny of every species is extinction. It seems that only a third of all organisms that have left a fossil record have representative species that are still living. This means that literally millions of species are now extinct, many of which were very successful in their day.

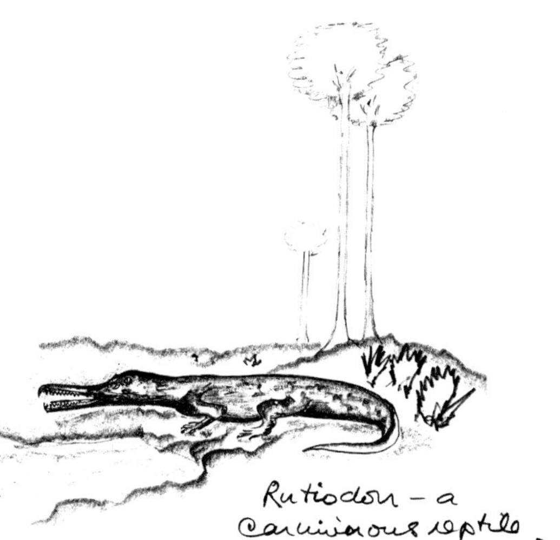
Rutiodon – a carnivorous reptile of the Chinle formation

At the end of the Devonian there was a severe decline in the numbers of trilobites, brachiopods, corals, fish, and many other animals. All temporarily recovered. At the end of the Permian, the decline was even more severe. The trilobites became extinct and the corals and the brachiopods lost their previous overwhelming dominance of the shallow marine world. By this time life on land had been well established but this too was severely devastated. The amphibians and reptiles were reduced to less than a quarter of their number. In fact the number of all known families of animals at the end of the Permian was subsequently reduced by half. At the end of the Cretaceous, by which time life had proliferated on Earth, a quarter of the animal orders, including the dinosaurs, became extinct.

There have been many theories to explain these periods of extinction. There appears to be no evidence to support sudden catastrophe of a kind that could have wiped out particular families of animals and not others in a single event of annihilating proportions. Nor does it seem likely that the devastation of life during a hiatus was generally prolonged over tens of millions of years, because the possible cause of the decline would have had time to register in the fossil record. Yet, although no one knows how long extinction took for any one species, the decimation of so large a number of species in any one episode seems likely to have been spread over an extended period of time.

Plants seem to be perpetual survivors of periods of mass extinctions, for there are comparatively few instances of major terminations of their lines of evolutionary development. If a supernova had bombarded the Earth's surface with sufficiently powerful doses of radiation somehow to cause the selective demise of so many animal families, plants should immediately and irrevocably have been affected, but their fossil record does not appear to show this scale of interruption. A reduction of trace elements in soil, the onset of devastating fungal or viral diseases, and other disasters that might have been selective, may very well be valid for continental localities but do not seem to have occurred on a planetary scale. Perhaps we are all expecting too sensational an explanation, when most of the answer is already available. Many paleontologists and paleobiologists now think that this may be the case.

The biosphere is in equilibrium. If part of it is altered, other parts will compensate, perhaps unfavorably for life as it was before the need. For instance, in some authoritative opinions the continued use of aerosols with fluorocarbon propellants may threaten the degree of protection that the ozone layer of the biosphere provides against ultraviolet radiation. It is thought by some that the propellant might gradually change the balance of the biosphere by causing such damage and therefore alter the character of life on Earth. Life would go on but it might become differently balanced. Any life form that could not adjust to increased ultraviolet radiation would eventually become extinct and be replaced by resurgent forms that had survived. But there would be no sudden catastrophe, although in a fossil record of the future it might appear that there had been.

The forces that caused the mass extinctions of the past are now thought likely to have been not changes in world climate alone nor the drifting of continents into different latitudes. Nor were they the result of interaction of merging coastal areas between colliding continents, nor the formation of mountain ranges, increased glaciation, or the changing pattern of currents in the oceans. Nor were they the consequence of the fluctuating levels of the sea, which influences the extent of continental sea incursion and regression. The suggestion is that all these dynamic forces worked in unison. This argument does not preclude an occasional cataclysmic happening; it simply suggests that such individual events at most contributed to an imbalance and were not its only cause. Although only the plants appear to have been comparatively little affected by events, in fact their distribution was very much affected by continental drift, and their evolution by the gradual environmental changes such movement encompassed. What the fossil record of the plants appears to demonstrate is that on a year-to-year basis through time there does not seem to have been a planetary cataclysm at any time since the rock record of land life began.

Periods of extinction benefited evolution by determining the survival of the most adaptive forms of life. This perhaps explains why each period of decline was followed by ever more vigorous development of the animals and plants that did survive. They were simply better equipped for life than their predecessors. At the end of the Cretaceous Period when the dinosaurs lost their dominance of the land, two such life forms burgeoned forth as if they had been waiting in the wings to take the stage in the present Cenozoic Era, which commenced sixty-five million years ago. The angiosperms, the flower-bearing plants that spread explosively into the quarter of a million or more species which populate Earth today, and the mammals.

Mammals arose from the mammal-like reptilian therapsids, which had continued to develop slowly through the time of the dinosaurs. Those that survived into the Cenozoic were quite small. Under twenty pounds in weight, they had developed a brain proportionately larger in size than the reptiles ever did, had a single lower jawbone where reptilian jaws had several elements, their hearing apparatus was more complex, and their teeth were differentiated into molars, canines, and incisors while reptilian teeth are not. There were many other anatomical differences between mammals and reptiles at the turn of the Cretaceous, differences which still distinguish them from other creatures today; but in addition to these differences, they were very much more

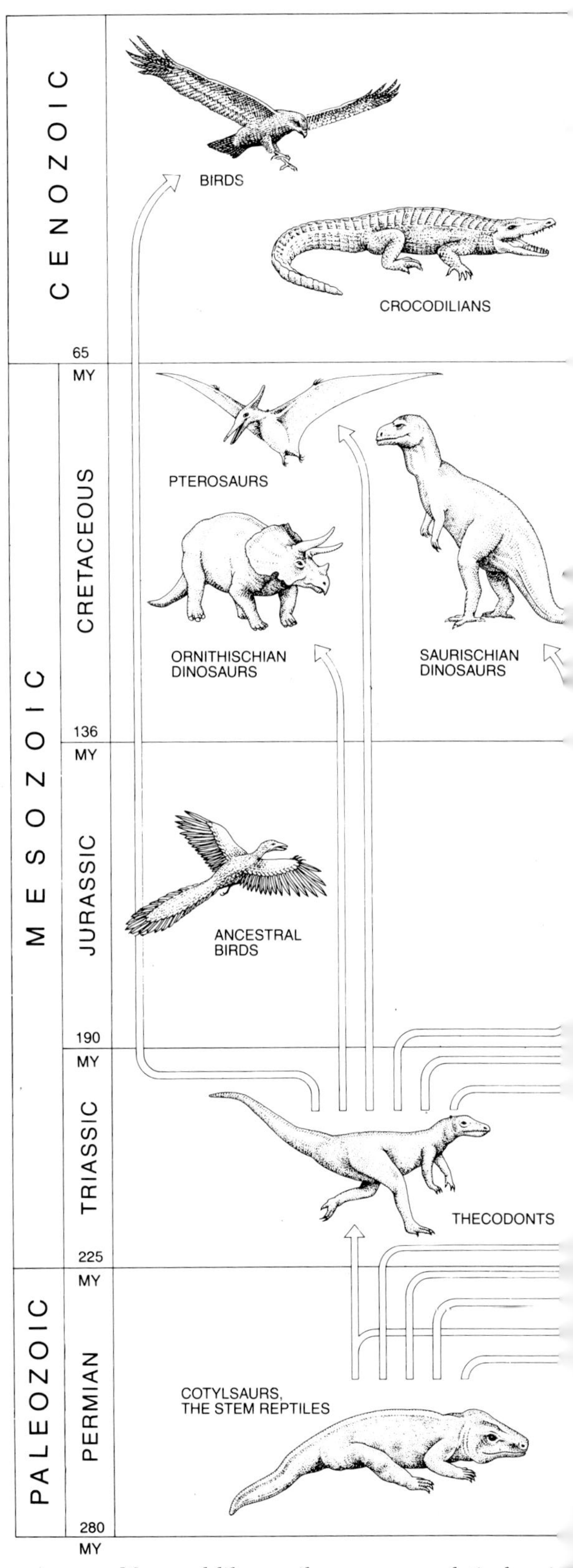

Mammal-like reptiles occupy a relatively mi- role in the entire reptilian line of descent. At lower right are the therapsids, with the ictidosaurs one of the reptilian

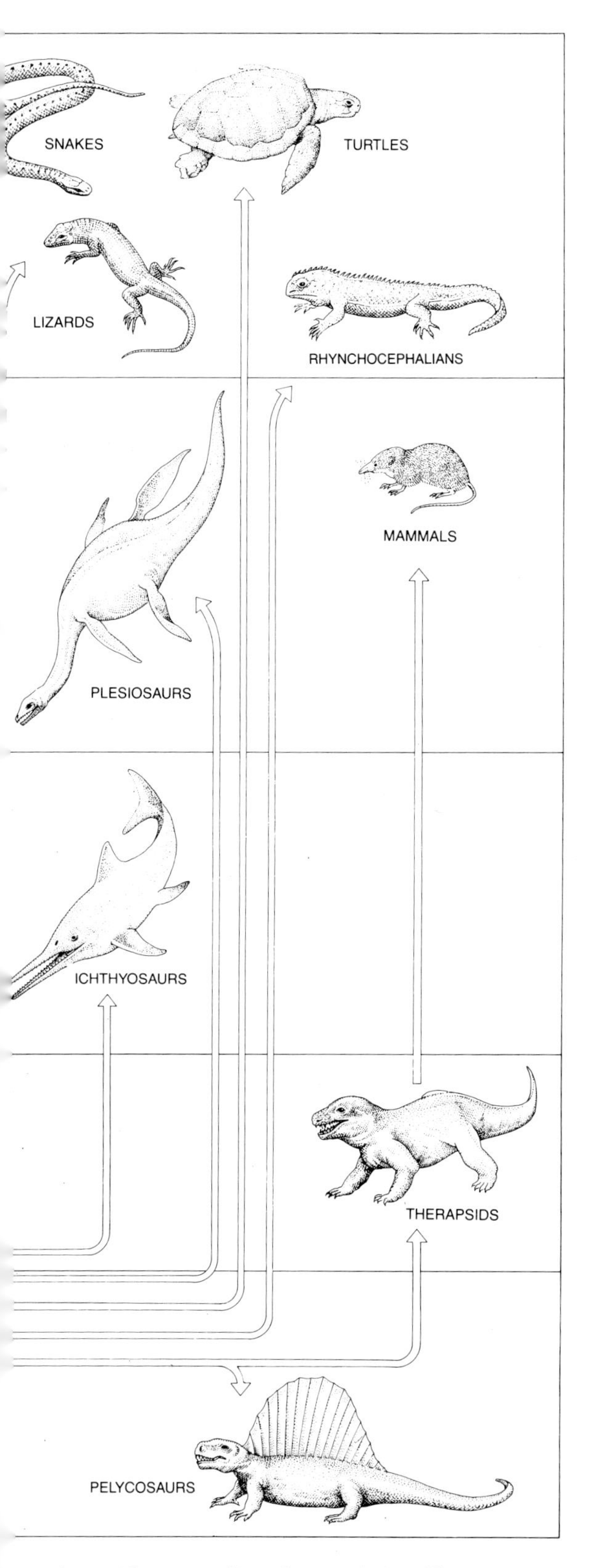

orders with mammalian characteristics. The early mammal, which is shown in the Cretaceous Period (near top), is rather similar to the present-day oppossum.

alert. It is not known at what stage the basic mammalian features of constant body temperature, reproduction, the suckling of infants, and the growth of protective hair first appeared in combination. Unfortunately the soft parts of an animal only fossilize in rare instances. But it is known that by this time—sixty-five million years ago—the first primates had evolved. They were small, tree-dwelling animals, the "plesiadapids," from which tree shrews and lemurs and ultimately all other primates evolved.

By the middle of Cretaceous times Pangea was well advanced in the process of splitting into the elemental continents (see illustration page 23). Some groups of continents were still firmly attached to each other, while others were only in tenuous contact. Whether by direct land routes or by isthmus, mammals spread to every part of the habitable world. During Cenozoic times the continents drifted into the relative positions they are in today. Fairly early in this process many animals had become isolated from each other on different continents. The root stock had become dispersed, but the processes of evolution continued. Animals, and particularly mammals, adapted to different ways of life to survive. They learned to fly, to walk, to jump, to climb, and to burrow. These adaptations happened simultaneously on different continents to animals from the same stock that found themselves in similar environments. Their independent evolution led to the development of very similar-looking and behaving animals in widely separated parts of the world, a process called "adaptive radiation." The opposite process whereby unrelated animals tended to become alike in their way of life as a consequence of their confinement in particular regions is called "adaptive convergence."

But in spite of the now mostly separated mainland continents some mammals were still able to disperse by traveling the attenuated necks of land that joined North America to Eurasia and to South America. One of these animals, the camel, evolved in North America during Cenozoic times. Only a few million years ago during the Pleistocene Epoch, it spread to Eurasia and Africa, and down the west coast of South America. A track of footprints of one of their kind is preserved in the sediments of Coalpits Wash at Zion. The camel was (and still is) a very strange animal. From its present-day descendants it is thought that in winter it could live for months without water and, having lost a quarter of its body weight in the process, it could replace all of it in ten minutes' drinking time. In summertime it could allow its body temperature to rise to one hundred and six degrees Fahrenheit before sweating. Its urine was concentrated to conserve moisture and the animal grew thick fur for insulation. The animals that made their way to South America evolved into llama, those in Eurasia became two-humped camels, and those that spread into North Africa became the single-hump camel of the Sahara. The species that left its track at Zion was probably the *Camelops*, a single-humped animal which, with all its kind, disappeared permanently from the North American scene about ten thousand years ago.

Throughout the recent history of the region of Grand Canyon, Zion, and Bryce, the same processes of evolution continued and are continuing today. The Kaibab tassel-eared squirrel, which is only found on the North Rim, has evolved differently from its more common relative, the Abert squirrel, which is found in the area of the South Rim and other parts of the Southwest (see illustration pages 148–149). Although they derive from common stock, the Kaibab squirrel became separated from its brethren by the Colorado River, the Grand Canyon, and adjacent low-lying desert country. Both squirrels are ecologically tied to the ponderosa pine forest, but the Kaibab squirrel belongs to a relatively small and confined community on the North Rim that permitted genetic drift. This means that mutations have been perpetuated by interbreeding on the North Rim instead of being dissipated, as they were on the South Rim. The result is squirrels of an entirely different coloration on opposite sides of the Canyon.

There are other examples of small but discernible differences that have developed between animals as a consequence of the Grand Canyon acting as an evolutionary barrier. They include species of ground squirrel and pocket mouse and subspecies of pocket gopher, chipmunk, and other mammals. In each instance it is those animals that have become isolated on the higher altitude North Rim of the Canyon that have probably modified from the common stock, to a greater extent than their relations on the South Rim. One adaptive change that has taken place that does not seem to relate

Grand Canyon as barrier—
Sunset from Hopi Point

This is a good place to get a sense of Grand Canyon distances. On the far horizon ninety miles away (center, left) are the volcanic mountains of the Toroweap region. The South Rim at left and the North Rim at right are separated by six miles and a difference in elevation of over a thousand feet. The Colorado River can be seen

shining in the evening light near Boucher Rapid a mile below the adjacent rims. Grand Canyon and the Colorado River act as a barrier to the wildlife (other than the birds) that lives in the Canyon. Some species, notably the Abert (South Rim) and the Kaibab (North Rim) tassel-eared squirrels, have evolved differently from a common stock because of this. 31. F-4.

to isolation but to environment alone is that of the pink Grand Canyon rattlesnake, which is found on both sides of the Colorado River and is unique to the Canyon.

Between the upper slopes of the San Francisco Peaks to the south, the high plateaus above Zion and Bryce canyons to the north, and the shore line of the Colorado River in the depths of Grand Canyon, there lies a spectrum of life equivalent to a two-thousand-mile change in latitude from north to south. The scene varies in its similarity to the delicate Arctic-Alpine and Hudsonian zone world of the Canadian tundra at one extreme, and the harsh, unforgiving world of the Lower Sonoran deserts of Southern California at the other. And life forms vary from the delicate alpine plants and kinglets of the high country to the resilient barrel cactus and scorpion near river level. As one reflects on the scene, it is difficult to avoid philosophizing. Could all this have evolved from nature's invention of the eukaryotic cell and the magic number—two?

The life-zones of Grand Canyon

The Grand Canyon teems with life. Varying ecosystems thrive in particular areas of sun, shadow, and exposure to certain conditions of wind and weather. The determining factor for such a community is usually its elevation. The life forms illustrated here broadly represent the series of zones of life found in the Canyon region and a range in geographical latitude from the desert lands of Baja California in the south to the Canadian border in the north. Grand Canyon formed a barrier between some animal species, and those on the North Rim were confined there because the river and surrounding desert obstructed migration. They also had to adapt to a higher altitude because the North Rim area is generally a thousand or more feet higher than the South Rim. Thus, the Abert and Kaibab squirrels are derived from common stock; the Abert is commonly found on the South Rim, but the Kaibab Squirrel on the North Rim is very different in appearance and is now a rare species. The Grand Canyon Rattlesnake is an example not of separation, because it is found on both sides of the Colorado River, but of adaptation to environment, for it is found only in Grand Canyon.

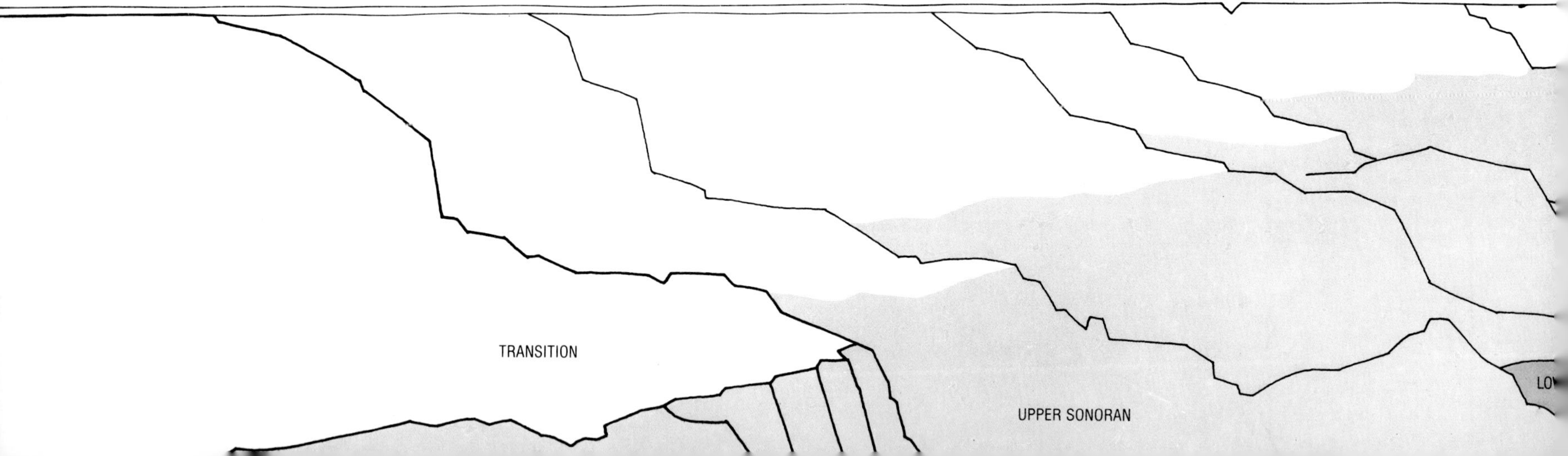

ENGELMANN'S SPRUCE
Transition, Canadian
MOUNTAIN LION (COUGAR)
Upper Sonoran, Transition
RAVEN
Upper Sonoran, Transition
PINYON JAY
Upper Sonoran
GAMBEL'S OAK
Transition
ARIZONA GRAY FOX
Upper Sonoran
LITTLE SPOTTED SKUNK
Lower Sonoran,
Upper Sonoran
WESTERN TANAGER
Transition
KAIBAB SQUIRREL
Transition
TARANTULA
Lower Sonoran
BARREL CACTUS
Lower Sonoran
GRAND CANYON
RATTLESNAKE (pink)
Lower Sonoran,
Upper Sonoran
INDIAN PAINT BRUSH
Upper Sonoran
WHIPTAIL LIZARD
Lower Sonoran
NORTH RIM
CANADIAN
TRANSITION
LOWER SONORAN
UPPER SONORAN
HSB

IX

INDIGENOUS MAN

Wupatki ruins

Built from Moenkopi Sandstone by the Sinagua Indians, Wupatki was occupied by them for nearly a century from A.D. 1120. Their dwelling here was by far the largest of about eight hundred such in the region. It had a ball court as well as a large amphitheater.
32. F-5.

The oldest known artifacts yet discovered in Grand Canyon are small figurines made from willow twigs of animals pierced by miniature spears. They were found under rude cairns in the deep recesses of caves in the Redwall Limestone, and it is believed that they were put there by American Indians as part of some hunting ritual at least four thousand years ago. The people who made them were probably members of what archaeologists call the "Desert Culture," one of several geographically defined groups of ancient peoples who populated North America at that time. Although this may sound ancient in modern historical terms, in fact these people were by no means the first Americans. Their original ancestors preceded them by many thousands of years.

Since the continent of Africa is generally held to have been the cradle of mankind, how did early *Homo sapiens* get to North America and into the depths of Grand Canyon? The story that answers this question starts in the year 1904 with the discovery of a biochemical method for classifying the relationship of mammals. George Henry Falkiner Nuttall was a distinguished biologist and physician who was born in San Francisco in 1862 and died in London in 1937, having acquired British nationality in 1900. Nuttall demonstrated that if blood serum from one animal is injected into another, antibodies will be manufactured by the latter—the immunological process. If the serum from the second animal is added to the serum of a third in a test tube, an insoluble solid, a precipitate, will separate out. Nuttall's hypothesis was that the genetic relationship of animal number three to animal number one may be judged by the amount of this precipitation: The greater the amount of this precipitation, the nearer the relationship of the animals to each other. Although Nuttall was able to corroborate his thesis, it was not generally accepted as a basis for genetic measurement for nearly seventy years. However, today scientists consider it among the most important developments in the study of human evolution, for it enables both objective and quantitative measurement of the relative distance between primates (or other mammals) on the evolutionary scale.

By using Nuttall's method, scientists have not only been able to confirm the shape of the evolutionary tree that has been reconstructed from the fossil record, but they also can measure the distance between branches on the tree and between twigs on the branches. They have shown that the evolutionary distance between man and the African apes is similar to distances between other related mammals. If the genetic distance of closely related animals is expressed as "one unit," the relative distance between unrelated mammals, expressed on the same scale, may be as many as seventeen units—man and whales for instance. But the distance between man and the chimpanzee is only about an eighth of one unit, and on this basis the chimpanzee is more closely related to man than any other animal. We share ninety-nine percent of the same genetic material with the chimpanzee, so it is believed that the evolutionary break between man and the apes happened quite recently. The fossil record suggests that this cannot have been less than four or five million years ago, and immunological evidence suggests that it is unlikely to have been more than ten million.

The earliest hominids probably bore a striking resemblance to the chimpanzee, an animal that walks using the knuckles of its hands for support while sometimes grasping a stick or other object. The chimpanzee can throw a stick quite accurately or use it to fish in a termite nest. It lives in a cooperative community, is primarily a ground dweller, and has a brain about a quarter of the size of modern man's—all of which fits the general description that anthropologists ascribe to early hominids. According to fossil finds in Africa, an advanced hominid had appeared by about three million years ago. Its particular form of pelvic bone allowed it to walk upright, and its skull had begun to change shape in keeping with an upright posture. This subhuman species had a brain about twice the size of its early ancestors and made simple implements, little more than chipped pebbles with sharp edges.

This willow figurine found in a Redwall cavern dates from at least 4,000 years ago.

Homo erectus, the first true man, appeared only about one and a half million years ago. Although his skull was still primitive, his body had further adapted to his now-customary upright stance and gait indicated by a pelvic bone that has changed little from that time down to the present. The brain size of *Homo erectus* was seventy to eighty percent of the brain size of *Homo sapiens neanderthalensis*, who did not appear until about a hundred thousand years ago and who had a skull that was almost modern in shape and structure. Man's brain has remained relatively unchanged in size since. *Homo erectus* made some beautifully shaped tools from whole pebbles, which are characterized by flakes cut from both sides. *Homo sapiens neanderthalensis* made tools from flakes of flint rather than the whole pebble. Modern man, *Homo sapiens*, did not appear until about fifty thousand years ago and is distinguished by a lighter skull, a slightly smaller brain, and an ability to make delicate tools and other objects, some of which appear to have been made for ceremonial purposes.

The ancestor of the hominids is believed by some anthropologists to have been *Ramapithecus,* an ape that lived in the Miocene Epoch, between five and twenty-two million years ago. During this epoch there was a broad land connection between Eurasia and Africa, which remained until continental drift caused the formation and flooding of the Mediterranean Basin about five million years ago. Sherwood L. Washburn of the University of California at Berkeley has suggested that this land connection may explain why fossilized remains of both *Ramapithecus* and some of modern man's now-extinct hominid relatives have been found in a wide area that includes the Middle East and the Balkans as well as Africa. The implications of this suggestion profoundly challenge the widely accepted view that modern man originated in Africa alone. And if Dr. Washburn's idea subsequently receives support from widely dispersed hominid fossil finds from continents in addition to Africa, then such confirmation will perhaps contribute to the explanation of the otherwise remarkably quick dissemination of mankind throughout the world, and the racial developments within the human species. Should such confirmatory evidence be unearthed it would indeed be a momentous contribution to the understanding of human ancestry.

No skeletal traces of *Homo erectus* or *Homo sapiens neanderthalensis* have been found in North America, but such traces of early *Homo sapiens* do occur. The earliest authenticated finds are those from a rich Pleistocene Epoch fossil bed at Old Crow Flats on the Porcupine River in northeastern Alaska. A fleshing tool made from caribou bone was found there in 1973, as were two mammoth bones that appear to have been deliberately crushed when green. They have been dated at about twenty-seven thousand years before the present (abbreviated to "B.P." in future references). In 1975 a cave that was occupied by prehistoric man was discovered near Avella in eastern Pennsylvania on the banks of Cross Creek, a tributary of the Ohio River. This site, now called the Meadowcroft Shelter, was continuously occupied from about twenty thousand years B.P., and by the dating of charcoal recovered from strata underlying those containing artifacts, it seems possible that the site could have been in use perhaps as early as thirty-seven thousand years B.P. Two other authenticated sites of a similar nature are those at the Fort Rock Caves, Oregon, and Wilson Butte Cave, Idaho. There are a number of other locations that lay claim to even greater antiquity, but although it is generally agreed by archaeologists that the date of early man's arrival in North America will probably be pushed back to perhaps fifty thousand years or more as new discoveries are made, only undisputed sites have been mentioned here. What is no longer in dispute is that North America became widely populated long before Europeans "discovered" it. It is generally thought that this was the result of spontaneous migration across what is now the fifty-six-mile gap of the Bering Strait.

During the last fifty to sixty thousand years there have been extended periods of time when a land bridge more than a thousand miles in width existed between northwestern Alaska and northeastern Siberia in the area of the comparatively narrow Strait. During what is known as the Wisconsin stage of the Pleistocene Epoch, ice sheets formed and glaciers advanced all over the Earth. Two huge ice sheets almost covered the whole of Canada and northern areas of the United States. In places these ice sheets were nearly two miles thick. On a global basis the amount of water locked up in the form of ice was so enormous that it substantially reduced sea level, which fluctuated according to the intensity of such glaciation. At one time during the Wisconsin stage of the Ice Age, sea level was reduced by more than three hundred feet,

and consequently the Bering Strait land bridge, called "Beringia," was open for about ten thousand years, approximately between twelve and twenty-two thousand years ago. Before this occasion, it is believed that there were three other extended periods of time during the Wisconsin stage when the Beringia was above the sea. The earliest of these, which could have influenced the migration of man from Asia to North America, is thought to have occurred around fifty to sixty thousand years ago.

One would expect that an ice age sufficient to cause sea regression would also be a time of year-round, severely cold climate in a locality as far north as Beringia. But apparently not only was this land bridge free from ice, but there were also considerable ice-free areas that extended deep into Siberia and into Alaska. Furthermore, there existed an unglaciated portion of Alaska that penetrated so deeply into its interior that it formed a corridor between the two great Canadian ice sheets via the region of present-day Edmonton, Alberta, and this corridor allowed access to the south. The corridor is called the "Alaskan Refuge," and the Old Crow Flats site of prehistoric man on the Porcupine River is a point where the corridor was sometimes blocked by advancing ice, temporarily preventing further migration.

The environment in the Alaskan Refuge was similar to present areas of the far north: grasses, lichen, mosses, and sedges, bright with flowers in the spring and the brief Alaskan summer, then the gray desolation, biting cold, and snow of winter. Beringia was open to animals as well as to man, and game was plentiful. Remains of caribou, musk ox, long-horned bison, and woolly mammoth testify to that. In fact, the motivation for human migration may well have been nomadic attachment to the wandering herds of animals who followed where the best food supply was available—no less so than their human predators. The people who came, the first Americans, were of Asian stock, and some anthropologists consider that they were Mongoloids and perhaps Caucasoids. But whatever else, they were tough—and their long journey southward ensured that only the fittest survived.

When they had traveled far south of the ice they found springs, streams, and lakes where playas and dried riverbeds are today. There were huge areas of spruce and pine forests and plentiful game—antelope, elk, deer—where few if any roam now. The west was characterized by lakes and swamps now gone, but the region north of the Colorado River called the "Great Basin" and the general area of the Colorado Plateau were much the same as they are today: sagebrush, juniper-pinyon communities, with lush stands of spruce, fir, and pine at higher elevations. It was a semidesert region with a few localities rich in food resources. By ten thousand years ago, so some anthropologists believe, the present-day distinctive American Indian attributes of stockiness, deeply tanned skins, dark eyes, and black hair seldom lost or grayed with age, had been established by normal genetic processes operating in isolation. The prehistoric men who made the willow-twig figurines in the Redwall caves of Grand Canyon were almost certainly of this description. Little is known about them beyond the probability that they were members of the Desert Culture and were the people who first settled, or at least regularly used, the region. They are known to have been hunters and gatherers of seeds, fruit, and other naturally occurring food. Figurines of a similar age and character to those that they placed in Grand Canyon caves have been found on both sides of the Canyon and beyond its rims. There is no evidence that people of the Desert Culture actually lived in the Canyon, but they may very well have done so.

The next human presence near Grand Canyon was not transient and tentative, but prolonged and evident. It lasted about six centuries, from about A.D. 600. These people lived in masonry dwellings organized in communities, and they and their descendants are called "Pueblo" Indians or "Puebloeans," after the Spanish word for "town." They used the plateaus and then the Canyon floor to support their farming activities. They are generally called the "Anasazi"—the old ones—of whom there were several groups, including the Kayenta and the Mesa Verde groups. There was also a much smaller group, the Cohonina, a different culture to the Puebloeans, living in the Grand Canyon region. When the Anasazi occupied the Canyon and its North and South Rims is known with some accuracy. Archaeologists relate the artifacts they find to particular strata. Then they cross-date them and compare them with other finds. Finally, because timber was often used in the construction of dwellings, they can date the building by using tree rings as a reference.

"Dendrochronology," as the technique of tree dating is called, was developed in 1904 by A.E. Douglass of the University of Arizona. By using his method, the date of a ruin often can be determined to within the year of its construction and sometimes to within the season of its construction. Tree-ring dating requires a section of a wooden beam used in the construction of an ancient building to be matched with the growth-ring pattern of long-lived trees of known age. In a sense, any tree's annual growth rings correspond to fingerprints, because the profile of successive rings is unique. Each series of consecutive rings is specific to the climate pattern of successive years, a profile that is never repeated exactly. A series of growth rings in a section of timber of unknown age can therefore be lined up with a matching sequence of growth rings in trees that have been dated. Dendrochronology is exacting, but it is an absolute method of dating, and is therefore an extremely valuable aid to historical reconstruction. It not only gives a date, but also the climate at that time.

Although Pueblo Indians began to live on Grand Canyon's rims at earlier dates, permanent farming sites did not appear on the South Rim until about A.D. 700 and on the North Rim until A.D. 850. Both the number and the size of these settlements steadily increased, but at this time the inner Canyon was only used for hunting, gathering, and experimental farming. Tree-ring evidence suggests that a slight change in climate by A.D. 1050 improved chances for crop survival in the Canyon, and it was then that the first permanent living quarters in Grand Canyon were established. As hundreds of ruins left by these ancient people testify, the population expanded until nearly all the arable spaces along the Colorado River, up side canyons, and on both rims, were in full use.

Two of the most populous places on the rims were at Tusayan, the ruin near the

Matkatamiba and Havasu

Grand Canyon hides many idyllic places. These two are among the most colorful, restful, and difficult of access: Matkatamiba Canyon at left and Havasu Creek at right. 33/34. E-4. 〈 〈

Nankoweap in Marble Canyon

An ancient dwelling and storehouse set high above the Colorado River and Nankoweap Rapids. It was occupied by Anasazi-culture Indians during the twelfth and thirteenth centuries. Then like thousands of other ruins all over the Southwestern United States, it was abandoned. The Anasazi were small and extremely nimble, as is attested by the lofty perch of this dwelling and the small size of its rooms. 35. E-4.

East Entrance to the National Park, and at an inaccessible place under a cliff near Point Sublime on the North Rim (see panorama pages 26–27). There are many ruins, such as that on Great Thumb Mesa, characterized by tower-like "fortifications," but although they look warlike, their true purpose is not known and probably was not defense. Down in the depths of the Canyon every Colorado River tributary delta was a potential living site. The best known are those at Nankoweap Creek in Marble Canyon about fifty-two miles below Lees Ferry (see panorama pages 156–157); at the confluence with the Little Colorado River; at Unkar Creek (see panorama pages 66–67, right-hand foreground); and at Bright Angel Creek near Phantom Ranch. Some dwellings were built on or near the farm sites, and their ruins can be plainly seen today. Others were built high above the river, under overhanging cliffs. Such cliff dwellings were also used as granaries to store surplus food for use in wintertime and are often difficult to locate because they are naturally camouflaged. Access to them is nearly always difficult or hazardous. Occasionally a wooden pole bridge, such as that near President Harding Rapid in Marble Canyon, can be seen joining two precarious ledges high above the river. They too were built by ancient Indians for reasons that are not always clear to those who have managed to reach them with the aid of ropes. Whatever else, these people were certainly superb rock climbers.

The only site that is still occupied in the Grand Canyon is at Havasu Canyon, seventy river miles below Bright Angel Creek (see panorama pages 160–161). All the others, both in the Canyon and on its rims, were abandoned between A.D. 1150 and A.D. 1200. Why the Canyon area became suddenly inhospitable or undesirable to its occupants is no longer a matter of conjecture. It is now generally agreed that years of drier climate simply reversed a process that seems to have started with a period of increased precipitation. Abandonment of Grand Canyon occurred many years before the "Great Drought" of the thirteenth century that is thought to have caused many Pueblo Indians to desert their communities in the Southwest. On the other hand, the Paiute, Shoshonian-related Indians, who first appeared on the North Rim of Grand Canyon and in the vicinity of Zion at that time, seem to have made use of at least some of the sites in the Canyon after their original owners had moved out—the site at Nankoweap is an example. Small groups of Paiute used the Canyon continuously until about 1885, when they were moved to their present reservation by the United States Government.

The people who live in Havasu Canyon are the Havasupai. They are descended from another culture called the Cerbat, which appeared in the Grand Canyon area about A.D. 1150, at about the time the Cohonina culture disappeared from the record. Havasu Canyon, on the south side of the Colorado River, is cut into the rocks of the terra-cotta-colored Supai Group and has a heavily mineralized stream of blue-green water running through it (see panorama page 155). The valley is a delight to the eye. Idyllic waterfalls, one more than a hundred feet high, cascade over stepped travertine formations (a cream-colored limestone) into emerald pools surrounded by lush grass and leafy glades under mostly beautiful skies. In springtime countless prickly-pear cactus blossom and wild bees buzz busily while pollinating the abundant flowers.

The Havasupai are a rotund, apparently easygoing and gregarious people. They are hunters above the rims of their canyon in wintertime, and farmers and fruit growers on the floor of their canyon in spring and summer. On hot summer afternoons temperatures are well in excess of a hundred degrees Fahrenheit. On such days, some women still gossip and weave baskets of intricate design while others join their men in the Havasupai equivalent of a club sauna bath. Rocks heated in a fire are placed in a central pit enclosed by a wooden frame covered with thatch. There are many such lodges beside Havasu Creek, and they are usually occupied by four people at a time for up to an hour. The Indians sing four rounds of a mythological song and then either dive into the creek or recover their strength on the bank before returning to the lodge, repeating the ritual four times. It has been suggested that this regular summer pastime is in some way responsible for limiting the Havasupai population to a maximum of about three hundred people, a number never exceeded in historic times—but high infant mortality is much more likely to have accounted for their small numbers.

As the Havasupai were establishing themselves in their canyon, another people, the Hopi (the Kayenta branch of the Anasazi) had already established Pueblo villages on the high mesas about one hundred and forty miles due east of Havasu Canyon on the far side of the Painted Desert. The Hopi people—an abbreviation of Hopituh, "the peaceful ones"—were joined in their villages by the Pueblo Indians who had abandoned Grand Canyon, a process which was complete by A.D. 1200. Hopi traditions are vague about their people's origin, merely suggesting that their ancestors climbed upward through three underworlds until they emerged into the present in an underground religious ceremonial chamber called a "kiva." Over the centuries and until comparatively recent times, the Hopi visited sites near the confluence of the Little Colorado and the Colorado River in Grand Canyon to collect the salt (used for religious purposes) that accumulates at the base of some rock formations. Ancient campsites have been excavated in these places and aboriginal potsherds have been found. Black-on-white and gray-colored examples of early Pueblo pottery were at the lowest level, and above them the later but still ancient yellow-ware, which is associated with Hopi pottery craft.

The first of a series of Hopi villages was established on high ground above the Painted Desert around A.D. 1200. Old Oraibi, as the village is now called, is possibly the oldest continuously inhabited place in the United States. Like all such settlements in desert areas, particularly sedentary ones with an agricultural economy, the prime consideration in the selection of the site was the availability of a dependable supply of water. In this case, more or less perpetual springs emerge from between the rocks that form the three principal mesas in the area. Today there are eleven Hopi villages in the group, including Old Oraibi, each a distinctly separate self-governing member of the Hopi community. A number of these villages, also of considerable age, are built in characteristic Pueblo style and are perched on one of the three mesas. They overlook the Painted Desert with a westward view dominated by the San Francisco Peaks on the far horizon.

Traditional Hopi religious beliefs reflect Hopi dependence on rain and an adequate crop from their fields. Their religious ceremonies therefore focus on their anxiety about fertility, germination, growth, and maturity of such crops. The Hopi believe that ancestral spirits continue to act as members of their communities and answer the prayers of the living by forming clouds that bring rain. These spirits, which include animals, plants, and other objects of the past, are called "kachinas." They carry

Havasu Canyon and the Esplanade (overleaf)

The Havasupai Indians have lived in Havasu Canyon for centuries. Here, from Boysag Point looking south, the Canyon can be seen in the humid early light of a summer's day in the right-hand area of the panorama. To the left of Havasu Canyon, standing proud of the wide platform of the Esplanade, Mount Sinyala rises lovely and serene. The Havasupai hold this butte to be sacred, and its name has been adopted by some of their most notable families.
36. E-3, 4.

prayers to the deities and directly influence the natural forces that sustain life on Earth. Kachinas are impersonated by members of the Hopi Kachina Society who, by putting on an appropriate mask and assuming the correct costume, reincarnate the spirit they represent. During the highly ritualized and colorful dances that are celebrated according to a strict calendar from the end of December to the middle of July, the dancers are manifestly the spirits they represent. Other religious ceremonies are performed during the remaining months of the year but the kachinas are no longer present, because the Hopi believe that between the summer and winter solstice the kachinas return to the San Francisco Peaks and then to their own spirit world.

On the other side of the Painted Desert from the Hopi Villages, west of the Grand Falls of the Little Colorado River a short distance northeast of the San Francisco Peaks, there are other reminders of the heritage of early man in the region of Grand Canyon. They are the ruined dwellings of a rather mysterious people called the Sinagua, who first lived in the area about A.D. 600 and disappeared from the scene six hundred years later. The Sinagua did not build their large community center at Wupatki until about A.D. 1120 and occupied it and nearby buildings for only a hundred years before deserting them (see panorama pages 150–151). Today the Wupatki ruin is a site of special interest because it not only had more than a hundred multi-storied rooms and a large well-built ceremonial arena in its vicinity, but also an equally large oval-shaped ball court. In southern Arizona and Mexico such courts are common, but this one is exceptional because it is constructed from masonry cleaved from Moenkopi Sandstone and not the more usual sun-hardened adobe clay-mortar. Little is known about the games men played here apart from the fact that they possibly used rubber balls, made from natural latex, of a type which have been found elsewhere in Arizona, Mexico, and Meso America. In its day Wupatki could well have been a meeting place for several cultures, for its location is known to have been near the limits of both Anasazi and Cohonina territory. There is little doubt that it was also the center for a local-league "ballgame" competition, because twenty courts similar in design to the Wupatki court have been found in the locality of the San Francisco Peaks. However, Wupatki was undoubtedly the Super Bowl of the Sinaguan league.

Volcanic eruption in A.D. 1066 made the large area east of San Francisco Peaks occupied by the Sinagua suddenly attractive. It was at this time that the Sunset Crater cinder cone was formed and much destruction wrought in its vicinity. Some of the Sinagua pit houses have been excavated from beneath the volcanic cinders and dust, a miniature Pompeii. Severe eruptions surely caused at least a temporary evacuation of the region. But later the Sinagua discovered an unexpected bonus in the ash-covered landscape. The land was now able to retain moisture after rain. Volcanic ash acted as an insulating blanket and dry mulch that improved the growth rate and quality of crops. These benefits attracted many new people to the area, and may quite well have influenced the decision to construct the Wupatki dwelling and others nearby. It follows that the increase in population might have strained land resources and that the more the ash cover was disturbed during cultivation, the less advantage it became as it was mixed with the topsoil. In due course falling yields and more mouths to feed may have contributed to the eventual abandonment of the area.

The Pueblo Indians discarded one place for another, sometimes after having established what appears to have been a large and balanced community numbering thousands of people over considerable areas. They lived in a delicate environment vulnerable to small changes in climate and did not have the technical skills to cope with change. They had seriously limited means of storing water, and springs simply dried up if underground water tables were not replenished by seasonal rain. A series of good years with surplus food, possibly accompanied by increased family, could be followed by years of drought.

One imagines the Puebloeans to have been peaceful, hardworking, and inventive. A number of the crafts that they developed survive to this day. The ruins they left, together with Hopi villages like Old Oraibi and Walpi, are monuments to their constructiveness. They never invented the wheel, nor understood the importance of the arch, but they mastered the art of dry farming, a skill that even modern farmers would find difficult to master. Then, around the year A.D. 1545, they were suddenly introduced to the ways of the modern world. The Spanish Conquistadores arrived to subjugate them. About sixty years later the Navajo Indians from Canada pillaged their homes, stole their women, and took their children into slavery. Later still, rebellion and war intruded into their lives. A new perspective was given to their anxieties. For the first time the world of indigenous man in Grand Canyon and the Painted Desert had been irrevocably changed by forces other than those of nature.

Walpi Pueblo

This picture of Walpi Pueblo, one of the Hopi Villages on the eastern fringe of the Painted Desert, was taken by John K. Hillers in 1879. Now the Hopi Indians do not permit photography on their reservation.

X

INTRUSIVE MAN

Redwall Cavern in Marble Canyon

This section of Grand Canyon is called "Marble Canyon" because the first explorer to go through was Major John Wesley Powell, who mistook the river-polished limestone (a sedimentary rock) of its walls for marble (which is metamorphic rock). The river makes several abrupt turns in this stretch which led to the undercutting of this huge cavern in the Redwall Limestone formation at the water's edge on an outside bend. Today, because floodwater is controlled by Glen Canyon Dam, the undercutting is at an end—at least for a while.
37. E-4, 5.

Trout sizzled gratifyingly in butter in the frying pan on the stove. Shrimp stewed slowly in their own juice. A favorite meal was on the way and a good book waited. My world closed in around me behind curtains drawn against the winter night. The motor home generator ran hard and the furnace full blast. The temperature outside had fallen to twenty below on the Fahrenheit scale. I had made an overnight stop on the northern edge of the Green Desert on my way from Denver, Colorado, to Bryce Canyon, Utah. Halfway to wilderness. The book was Elspeth Huxley's *Scott of the Antarctic*, a biography of a boyhood hero of mine. Here was a fascinating study of a man whom I had felt that I knew but soon learned that I didn't know at all, and of Wilson, one of his companions in that tiny tent through the Antarctic night, the artist, naturalist, and discoverer of the fossil *Glossopteris* on the Beardmore Glacier. Why men explore I felt I understood. But what is the force that drew these men back to that hostile place? How, I wondered, have other men looked at other intractable places such as the region of the Colorado Plateau, and what compelled them to return time and time again?

Carved on a rock at El Morro, near Gallup, New Mexico, there is an inscription in Spanish that says, "Here was the General Don Diego de Vargas, who conquered for our Holy Faith and for the Royal Crown, all of New Mexico at his own expense, year of 1692." That one sentence defines the motives of the Spanish Conquistadores. Their objective was to conquer, to pillage, and to enslave in the name of Christianity, and then to enjoy the patronage of the Spanish Royal Court. They were determined to find treasure and to subjugate those who possessed it, but they were singularly unsuccessful

Baron von Egloffstein: The Diamond Creek Region of Lower Granite Gorge, Grand Canyon.

Canyon perspectives

When modern artists portray the Grand Canyon they are disciplined by the thought that their subject is already well known. The artists of the nineteenth century were not inhibited in this way, and their work therefore tells as much about themselves as it reflects the character of the landscape. Baron von Egloffstein was one of the very first artists to visit the Canyon as a member of the Ives Expedition, which steamed up the Colorado River in 1857–58. His interpretation was strongly Wagnerian and not even the kindest view of it could claim resemblance to reality. William Henry Holmes and Thomas Moran, who worked in the region from the 1870's on, were artists of outstanding talent and integrity. Holmes portrayed the Canyon's intricate detail as if every rock was etched in his memory. He conveyed the geology of the place with uncanny accuracy, and the scale of it by using a matchless technique of panoramic drawing. Holmes often constructed his panoramas by combining sketches made from several viewpoints. In this way he could include features that could not be seen from one position. He balanced the composition of his drawings by using material from his field notebooks to portray otherwise nonexistent detail in the foreground or at the extremities of each picture. Holmes became a distinguished anthropologist, and in later life was appointed director of the National Gallery of Art in Washington. His panoramas are unique in the annals of technical drawing. Moran was a student of J.M.W. Turner, one of the great landscapists. There is a strong hint of Turner's impressionist influence in Moran's paintings of Grand Canyon; they avoid detail in favor of mood and character.

W.H. Holmes: Two drawings from Vulcan's Throne—at left, below, the westward aspect, to the volcanic region of Mt. Trumbull. At right and to the north, Toroweap Valley. Topographically continuous, these look like two parts of a panorama, but in fact were not so intended.

Thomas Moran: Sketch for painting, "The Transept."

W.H. Holmes' drawings "Portals of the Virgen" (top, left) and "Panorama from Point Sublime" (bottom) were both used to illustrate (with Moran's "The Transept") Clarence E. Dutton's Tertiary History of the Grand Canyon District *(1882). The third picture (top, right) is of the view from Vulcan's Throne looking due east across the Esplanade with the majestic cliffs of Toroweap in the middle foreground to the right. The reader should compare these illustrations with the panoramas on pages 26–27, 30–31, and 104–105.*

Bryce Amphitheater

At any time of year Bryce Canyon is an extraordinary spectacle of light, color, form, and mood. For the photographer it has one particularly pleasing quality—it is almost impossible not to get good results. Unlike Grand Canyon or Zion, the scale is relatively compact, the scenery is easily accessible, and the reflected light combines with high elevation (8,000 to 9,000 feet) to ensure detail in the shadows even in the brightest light. Bryce challenges the serious photographer's ability to compose a scene, convey a mood, and, in the author's case, to illuminate the character of the Amphitheater as a whole in both winter and summer. The 35mm inset picture was taken with a 50mm lens from approximately the same position in summertime as the winter panorama on this page. Its purpose is to illustrate the difference in coverage using the author's panoramic technique in comparison with more conventional pictures. The panorama on the next page is of Bryce Amphitheater in summer, but taken from the heart of the Canyon on the Peek-a-Boo Trail. The panorama covers 100 vertical and 200 horizontal degrees.

38/39. C-4.

Mather Point

This is one of the most popular viewpoints on the South Rim of Grand Canyon. The panorama was taken mainly to convey both the scale of the Canyon and the precipitous nature of its flanks. You are looking almost vertically downward into the depth of the canyon a mile below. The 35mm inset picture was taken from the same location with a 50mm lens for comparison. 40. F-4.

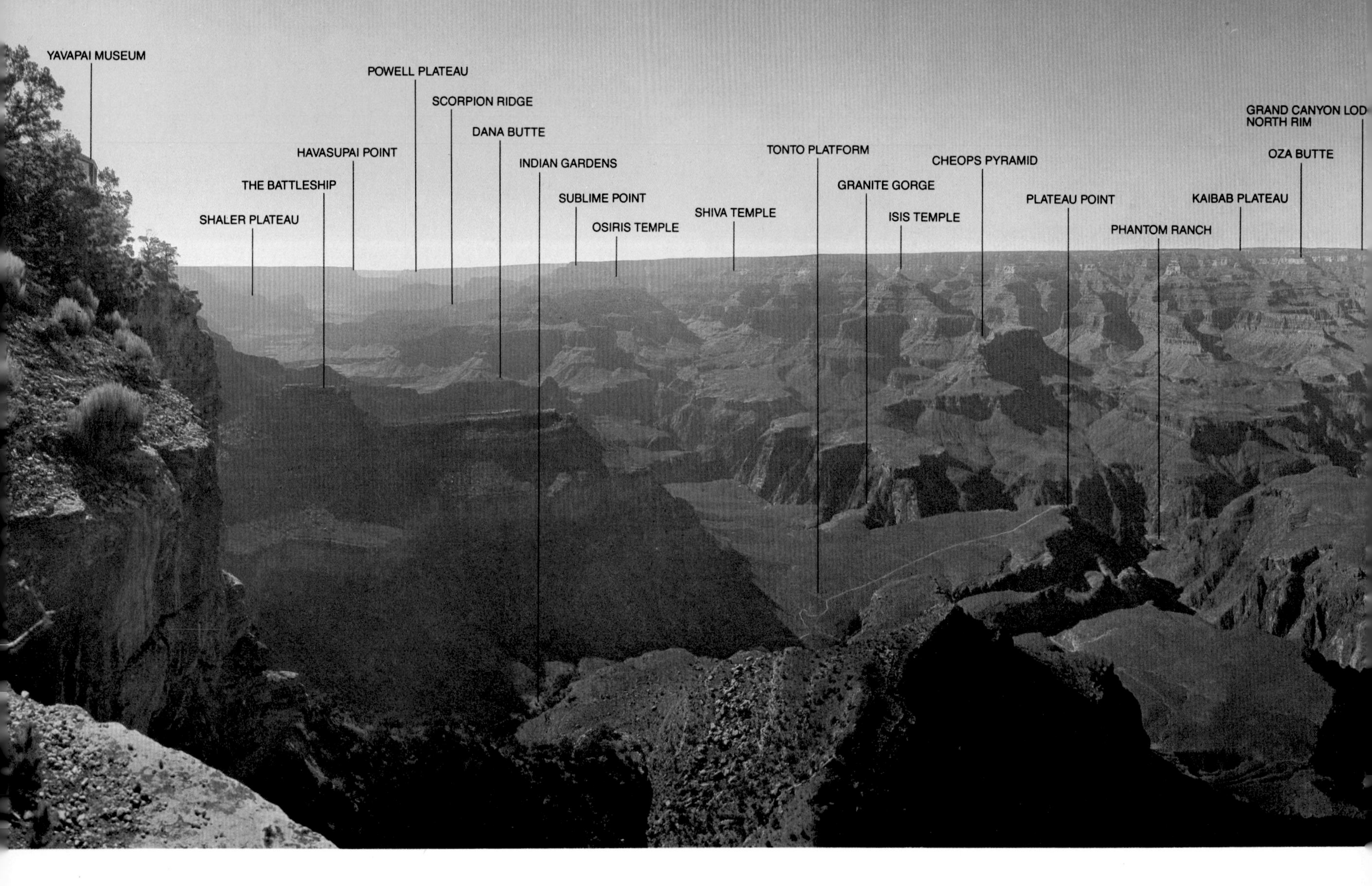
YAVAPAI MUSEUM
POWELL PLATEAU
SCORPION RIDGE
DANA BUTTE
HAVASUPAI POINT
INDIAN GARDENS
THE BATTLESHIP
SUBLIME POINT
SHALER PLATEAU
OSIRIS TEMPLE
SHIVA TEMPLE
TONTO PLATFORM
GRANITE GORGE
ISIS TEMPLE
CHEOPS PYRAMID
PLATEAU POINT
PHANTOM RANCH
KAIBAB PLATEAU
OZA BUTTE
GRAND CANYON LOD
NORTH RIM

The view from Yavapai Museum

Yavapai Museum is perched on the edge of a precipitous drop on the South Rim of Grand Canyon, near Grand Canyon Village. Yavapai is one of the most popular overlooks for the several million people who visit Grand Canyon National Park every year. The museum not only houses an interesting exhibit that tells the geological story of Grand Canyon, but also has a multipanel picture window through which visitors can identify the mountainous buttes that rise in tiers from the Canyon's floor directly below. This panorama reconstructs that view. 41. F-4.

on the Colorado Plateau. The first Europeans to see the Grand Canyon came to verify a fanciful tale by Álvar Núñez Cabeza de Vaca, one of a group of four men, including a Negro named Esteban, which was the surviving remnant of a Spanish expeditionary force that had set out from Florida in 1528, thirty-six years after Columbus had discovered America. After years of successive disasters and incredible endurance, the quartet was reunited with Spanish forces in New Spain (as Mexico was then called), in 1536. Cabeza de Vaca reported to Antonio de Mendoza, the viceroy of New Spain, the existence of great cities of fabulous wealth to the north. In 1539 Mendoza dispatched a priest as an envoy, Fray Marcos de Niza, who was preceded by Esteban, to find the cities and plunder their treasure.

On his return to New Spain, de Niza reported quite untruthfully that he had seen seven golden cities in the distance. He was referring in part to Hawikwah, a striking but unmistakably adobe village of central New Mexico where Esteban had died at the hands of the Pueblo residents (the Zuni) and where de Niza learned that the Indians called the locality "Cibola." On the strength of the doubtful intelligence supplied by de Niza, Viceroy Mendoza equipped two much larger expeditions to find and plunder the mythical "Seven Cities of Cibola." One of these expeditions was led by Francisco Vasquez de Coronado, who reached Hawikwah in July 1540 to find irrefutable adobe Pueblo dwellings in place of the fabled golden-paved streets. Coronado then learned of a Pueblo town called Tusayan (now Old Oraibi and not to be confused with the South Rim ruin by the same name) to the northwest. He dispatched his able lieutenant Pedro de Tovar to investigate. Having found Tusayan to be just another disappointing adobe Pueblo village, Tovar could only report unfavorably to Coronado. Tovar had mentioned in his report that on his travels he had heard of another Indian tribe (presumably the Havasupai) who lived near a great river far to the west. Undeterred by past failure, Coronado promptly dispatched yet another lieutenant, García López de Cárdenas, to find these unknown people with the intention, no doubt, of relieving them of their golden treasure.

Guided by Puebloeans from Tusayan, Cárdenas and his men reached the vicinity of the South Rim of Grand Canyon after a twenty-day march in September 1540. The point on the South Rim from which the Cárdenas party first glimpsed the Canyon is unknown but the popular belief is that it was probably within a short distance of present-day Desert View near the East Entrance to the National Park. Since there is no warning glimpse of the Canyon when approaching it from any direction, the total unexpectedness of the chasm is maintained until literally the edge is reached. One must believe that the Cárdenas party responded to the sight that suddenly confronted them with at least a gasp of astonishment and disbelief. They certainly misjudged the Canyon's scale for Cárdenas reported that after three days of effort to descend to river level such descent not only proved to be far more difficult than they had expected, but the three men who got part of the way down found that, as they were approached, rocks that had looked small from the rim grew substantially in apparent size. On the fourth day the Cárdenas party, now short of water, started the return journey to Tusayan. The myth of Cibola seems to have died at this time and over two centuries passed before Grand Canyon was again viewed by European eyes—by Fray Francisco Tomás Garces who visited the Havasupai in A.D. 1776. But the laugh on Cárdenas was with the Puebloean guides, the Hopi. They not only knew several trails down into the Canyon which obviously they did not reveal to Cárdenas, but in all probability they led the Spaniards on a merry caper around the region, possibly to avoid introducing the intruders to their centuries-old trading customers, the Havasupai. How else could the party have taken the reported twenty days to complete a journey from Old Oraibi to Desert View, which should have taken no more than three or four?

In 1680, one hundred and forty years after the Conquistadores first saw Grand Canyon, the Puebloeans successfully rebelled against the Spanish. The Spaniards' counterattack came in 1692 and took a subtle form. General Don Diego de Vargas, whose expedition the reader will recall was "at his own expense," enlisted one Pueblo community against another until most of the Puebloeans were subjugated. The Hopi, apart from contracting European diseases to which they had no immunity, remained free of the Spanish yoke. Meanwhile, their villages were disturbed by the appearance of another threat, the Navajo.

The Navajo Indians first appeared in New Mexico about 1550. They moved westward, raiding Hopi pueblos and stealing what they could, particularly children for slavery and women for concubinage. Their propensity for Hopi captives and their contact with other Pueblo Indians profoundly influenced their own way of life. They slowly adapted Hopi religious ceremonies. They learned to tend sheep and goats (introduced by the Spanish), to till the soil, to weave, to make and paint pottery, and to trade. The Navajo did not adopt the Pueblo village but preferred widely separated hogans—dome-shaped and adobe-covered dwellings, supported by wooden frames. They usually had a winter and spring home on the desert floor, and a summer hogan on a high mesa. But while they successfully adapted to more sedentary ways, they also retained their enthusiasm for marauding.

By the time English-speaking Americans began to permeate the Four Corners region after its secession to the United States as part reparation following Mexico's defeat in the War of 1846, the Navajo were already herdsmen, weavers, and farmers. They rode horses, no doubt taken from the Spanish who had reintroduced the horse to North America centuries before, and traded their tribal wares for metal goods, saddlery, and firearms. They were easily provoked, and routinely raided other tribes or attacked those they considered usurpers. With comparative mobility and local superiority of arms, their raids were just too successful for the comfort and well-being of people who were within their range of activity.

During the American Civil War both the Navajo, and the Apache farther to the south, took great advantage of the preoccupation of the military to raid at will. Following unsuccessful attempts by government officers to negotiate peace with Navajo "tribal chiefs" (it was not understood that in reality the Navajo representatives had little tribal jurisdiction), and after a further series of skirmishes and raids, Colonel Kit Carson was instructed to enter Navajo country in June 1863. What followed caused a depth of bitterness that survives to this day.

Carson and his men attacked and put the Navajo to rout. Their orders were to destroy all Navajo crops and livestock. Many Navajo were killed in the process. Some escaped, and those who remained were made captive. They were forced to quit the desert land in which they had settled and learned to live. They were rounded up and force-marched three hundred miles across mostly barren country to the alien territory

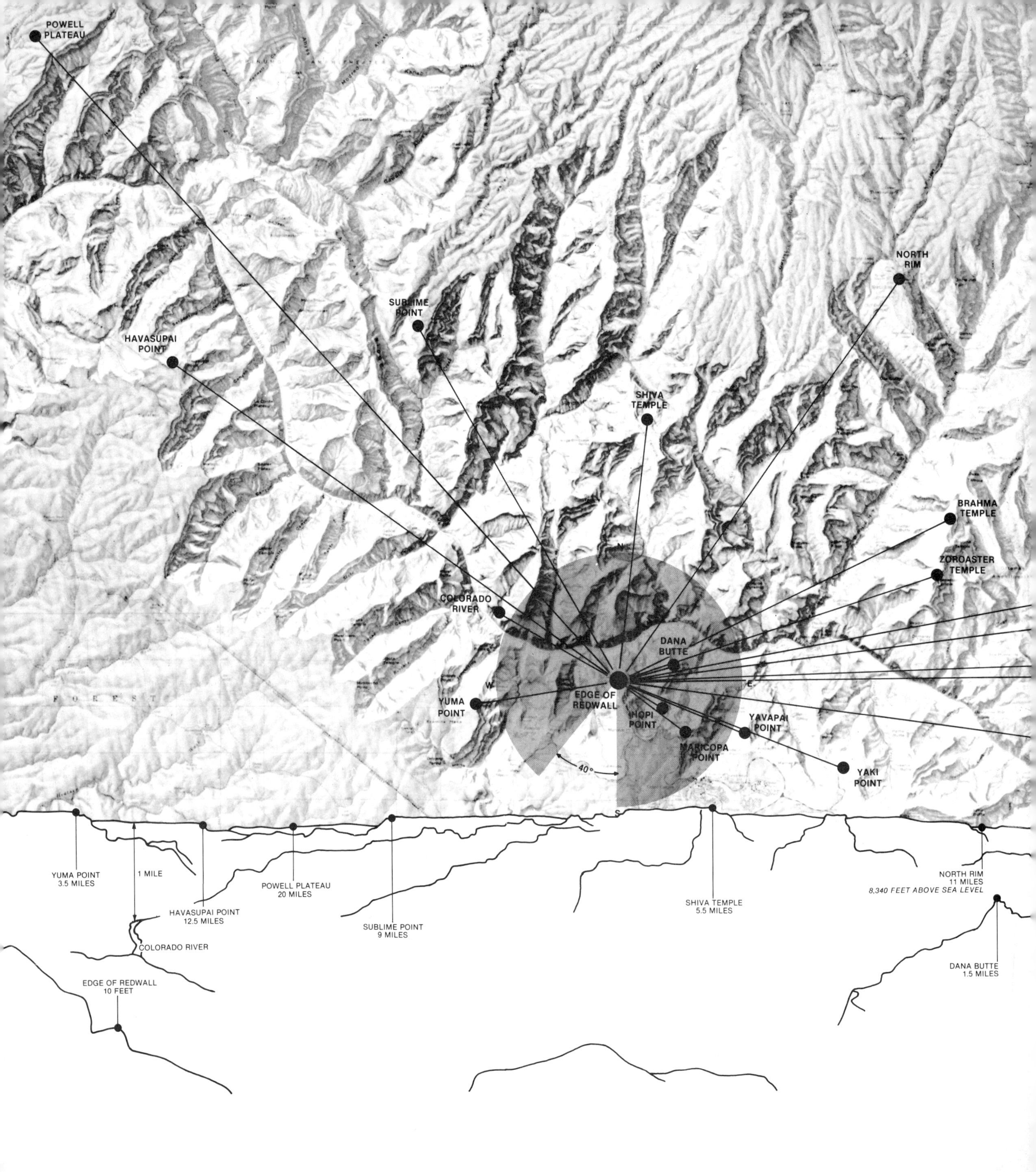

POWELL PLATEAU
NORTH RIM
SUBLIME POINT
HAVASUPAI POINT
SHIVA TEMPLE
BRAHMA TEMPLE
ZOROASTER TEMPLE
COLORADO RIVER
DANA BUTTE
YUMA POINT
EDGE OF REDWALL
HOPI POINT
YAVAPAI POINT
MARICOPA POINT
YAKI POINT
40°
N
S
E
W
YUMA POINT 3.5 MILES
1 MILE
HAVASUPAI POINT 12.5 MILES
POWELL PLATEAU 20 MILES
SUBLIME POINT 9 MILES
SHIVA TEMPLE 5.5 MILES
NORTH RIM 11 MILES
8,340 FEET ABOVE SEA LEVEL
COLORADO RIVER
EDGE OF REDWALL 10 FEET
DANA BUTTE 1.5 MILES

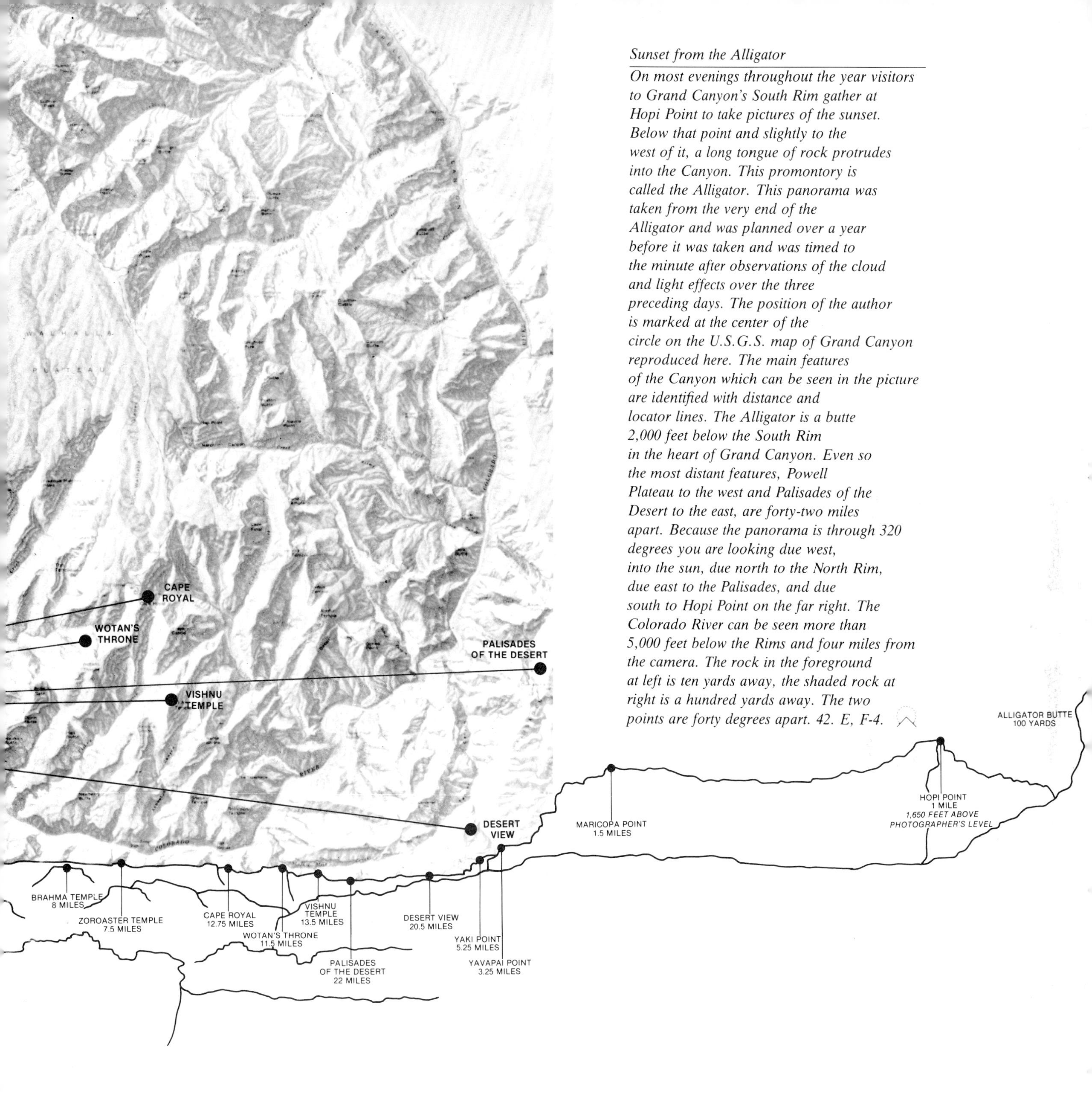

Sunset from the Alligator

On most evenings throughout the year visitors to Grand Canyon's South Rim gather at Hopi Point to take pictures of the sunset. Below that point and slightly to the west of it, a long tongue of rock protrudes into the Canyon. This promontory is called the Alligator. This panorama was taken from the very end of the Alligator and was planned over a year before it was taken and was timed to the minute after observations of the cloud and light effects over the three preceding days. The position of the author is marked at the center of the circle on the U.S.G.S. map of Grand Canyon reproduced here. The main features of the Canyon which can be seen in the picture are identified with distance and locator lines. The Alligator is a butte 2,000 feet below the South Rim in the heart of Grand Canyon. Even so the most distant features, Powell Plateau to the west and Palisades of the Desert to the east, are forty-two miles apart. Because the panorama is through 320 degrees you are looking due west, into the sun, due north to the North Rim, due east to the Palisades, and due south to Hopi Point on the far right. The Colorado River can be seen more than 5,000 feet below the Rims and four miles from the camera. The rock in the foreground at left is ten yards away, the shaded rock at right is a hundred yards away. The two points are forty degrees apart. 42. E, F-4.

of Fort Sumner, southeast of Santa Fe in New Mexico. This was the "Long Walk," as the Navajo call the savage hurt they suffered to their families, to their possessions, and to their way of life. Thousands of Navajo were herded into separate parties for the journey. About eight thousand survived the trek to Fort Sumner and remained incarcerated there until late in 1868. They received another and deeper injury—to their pride. They considered themselves to be *diné*—The People—a name the Navajo still use. When they were finally allowed to return to the region of Four Corners, they were a vanquished and broken people. The Navajo were impoverished and entirely dependent on charity to restart their lives in the malevolent desert. Those that had escaped the Armageddon of 1863 fled to the canyonlands neighboring their territory. One of their principal places of refuge was the Grand Canyon.

The Pueblo Indians, the Spanish, and the Navajo saw the Canyon in a very practical way; to them it was an ancestral home, a barrier and a refuge. In 1869, while the Navajo were resettling the desert area to the east of the Colorado River, a thirty-five-year-old white American, Major John Wesley Powell, and eight companions mostly in their twenties (three of whom died after leaving the expedition before its completion) were making the first traverse of the Colorado River through Grand Canyon. Powell went out of curiosity and adventure; he saw the Canyon and the people who lived near it intellectually, as a challenge to his understanding. Powell went into the Canyon again with a second expedition in 1871–1872, this time under federal auspices, and was destined to be involved directly or indirectly with the Colorado Plateau for the rest of his life. He consistently inspired others more expert than himself to probe, to sift, to think, and to offer conclusions about this natural wonder of the world. Eventually he became the second director of the U.S. Geological Survey and the head of the Bureau of Ethnology.

My first acquaintance with Powell was through his book, *The Exploration of the Colorado River and Its Canyons*, which I purchased at the time of the centennial celebration of his 1869 voyage. Powell wrote the text in the present tense and referred to places I couldn't find on the map because the names are no longer used. In the years that followed I gradually became aware of a great many inaccuracies and contradictions in Powell's account. One historian whom I met credited Powell's administrative abilities but acknowledged few others. This all appeared to contradict the reputation that posterity seemed to have bestowed on the man. What was Powell really like?

Major Powell was short, stocky, barrel-chested, and had an untidy, bushy beard. He was inclined to be beetle-browed; his nose was prominent, straight in profile but spatulate in shape. His piercing eyes were surmounted by a broad forehead with a vee-shaped hairline—all characteristics of his Celtic antecedents. Powell's most distinctive feature was the nearly elbow-length stump of his right arm. His forearm had been shattered by a minié ball fired by a rifle on the battlefield of Shiloh in April 1862. This was no ordinary bullet wound. When fired, the minié ball was expanded into an ugly cusp by the explosive it contained, causing absolute agony to anyone at the receiving end of its flight. Powell suffered the mental anguish of having to learn to write with an unaccustomed left hand and later of coping with the distress and pain of a phantom limb. The very fact that Powell had several resections of the stump indicates that repeated surgery did not improve his unhappy lot.

By nature Powell was an energetic man. The probability is strong that as a young man, nagging pain occasionally shortened his temper and perhaps warped his judgment of the moment. Subconscious compensation for his impediment might well have included a doubling of nervous energy, which added to his drive and aggravated his impetuosity. During the rigorous difficulties of his 1869 expedition the absence of his right arm must have tried his patience by making him too dependent on his companions, and must also have made note-taking a burden, which probably was why his log of the voyage was inadequate and he had to commit too many details to memory.

Powell had literally to be given an ultimatum by an appropriations committee of the House of Representatives to be persuaded to write an official account of his expeditions for government publication in 1875. As Powell states in the clearest terms in the preface to his popular version, *The Exploration of the Colorado River and Its Canyons*, it was never his intention to write the story of his adventures, but continuous public pressure upon him led to its publication in 1885—more than twenty years after the events portrayed in the book. But apart from lapses of memory, why the gross inaccuracies, the combination of two river expeditions into one, and the lithographs that portrayed part imaginary events and scenery? In his text Powell said of Andy Hall, a member of his first expedition, that he could "tell a good story and is never encumbered by unnecessary scruples in giving to his narrative those embellishments which help to make a story complete." A century ago people greatly depended on the written and spoken word for their entertainment, and telling a good yarn was part of that entertainment. I believe that Powell's approach to a popular (not scientific) book was to tell a good story based on fact but "unencumbered by unnecessary scruples."

In the latter half of the nineteenth century there was considerable public curiosity about the Colorado Plateau region. Until the Powell river expeditions of 1869 and 1871–1872 this area was virtually a blank space on the map of North America. On the maps of the day, the Green River disappeared southward into the desert lands of Utah; the Grand, as the upper Colorado River was called at that time, flowed southwest. Where the two great rivers met, and the character of the surrounding country for thousands of square miles, was mainly a matter for conjecture. But the existence of a great canyon complex to the west was known. Public interest had initially been aroused by the experiences of an expedition led by Lieutenant Joseph Christmas Ives in 1858.

In the wake of the Mexican War, Lt. Ives had been directed to report on the feasibility and possible extent of steamboat navigation up the Colorado River from the Gulf of California. He and his party, which included two artists and a geologist, voyaged upriver on the paddle steamer *Explorer*, which had been reassembled from parts shipped from Philadelphia. The journey ended abruptly against a rock in the river about twenty-one miles below the present Hoover Dam. Ives and his party were able to return partway downstream to rendezvous with a troop of men and mules and then to

Noonday rest

This vertical panorama was taken at the same place as the lithograph opposite. Note the difference in the height of the Colorado River—at least ten feet lower today as a result of the controlled flow at Glen Canyon Dam. Also note the exaggeration of perspective in the lithograph so as to impress the viewer with the depth and verticality. The lens used for the panorama was a "normal" focal length, so that the scene pictured is almost the same as that seen by the human eye. 43. E-4.

cut across country to approach Grand Canyon from the southwest via Peach Springs and Diamond Creek. According to Ives, when his party arrived at the river's edge in Grand Canyon he had the sensation that he was being hemmed in by walls two thousand feet high. He referred to the "depth and gloom of gaping chasms which imparted an unearthly character to portals of what might well have been infernal regions." In his report to his superiors he spoke of "harsh screams issuing from aerial recesses in the cañon sides and apparitions of goblin-like figures perched in the rifts and hollows of the impending cliffs, gave an odd reality to this impression." In his drawings of the scene one of the artists, F.W. von Egloffstein, portrayed exactly Ives' described Wagnerian impression of the Canyon. The net result of Ives' report and Egloffstein's pictures was to give the public an intriguing but completely erroneous impression of the Canyon. Ives rounded off his already inaccurate comments with the statement that "the region is, of course, altogether valueless. It can be approached only from the south, and after entering it there is nothing to do but leave. Ours has been the first, and will doubtless be the last, party of whites to visit this profitless locality."

Powell's first river traverse of Grand Canyon was headline news in 1869. The historic achievement not only filled some of the blank spaces on the map by locating the confluence of the Green and the Grand rivers, but also aroused national interest in

Morning run

Martin Litton's Grand Canyon dories lining up to run Upset Rapid, at Mile 150. Rated up to 8 on a scale of difficulty up to 10 because of low water, the rapid was at its maximum force.
44. E-4.

exploiting the incredible canyons themselves. Soon came the prospectors for minerals and precious metals; gold was found at Kanab Creek in 1871 and silver, carbonate of lead, and vanadium were reported in Havasu Canyon in 1873. Prospecting and mining activities began in the 1880's (and have continued on a modest scale down to the present). Mining pioneer Seth B. Tanner helped to reconstruct an old Indian trail into the Canyon from the South Rim, now called the Tanner Trail. William Wallace Bass, another prospector, established camp near Havasupai Point on the South Rim and constructed a cable-car crossing to an asbestos mine on the north side of the river. Nearby "Bass Rapid" perpetuates his name. Another asbestos miner, John Hance, began to supply visitors to the South Rim of the Canyon with log cabin accommodation and guided tours into the abyss while regaling them with tall stories for which he became famous. He gave his name to the Hance Trail and Hance Rapid. But most of the place-names in the Canyon were given by Powell and other members of the Geological Survey. "Grand Canyon" itself was officially so named by Powell after his first expedition.

Zion and Bryce canyons were in Utah territory, and Mormon settlers pioneered those areas during the 1860's. Flagstaff, Arizona, was founded in 1876, the same year as the Santa Fe Railroad reached the area. As a consequence of this, the number of visitors to Grand Canyon increased. Other prospectors followed Hance's example and found precious metal easier to extract from visitors than mineral wealth from Grand Canyon. Their exploitations aroused public concern for the fragile ecology of the area. The Canyon and its immediate vicinity was made first a Game Reserve in 1906 and then a National Park in 1919. Since that time the Park area has been increased on a number of occasions to include large regions of the North Rim and Marble Canyon. For the same reasons of conservation, Zion was also created a National Park in 1919, and Bryce Canyon in 1928.

Meanwhile scientific work in these spectacular places continued. By contrast with the prospectors, this work was in the hands of enlightened and far-seeing men who were interested in facts, not myths, and conveying images to the public which were significant impressions, not sensational ploys. Powell was certainly the man of greatest influence in this respect. Clarence Edward Dutton, seven years his junior, was Powell's willing collaborator. Dutton was a geologist. He joined Powell in the U.S. Geological Survey of the Rocky Mountain region in 1875 and spent the next ten years investigating the volcanics, the uplifting, sinking, twisting, and folding of the Plateau Province and nearby regions. In later years he was partly responsible for the development of the theory of isostasy—the general equilibrium of the Earth's crust. Dutton's classic report on the U.S. Geological Survey of 1880–1881, which he called *The Tertiary History of the Grand Canyon District* (an interpretation of the geology of land forms of the Grand Canyon), combined scientific detachment with perceptive romanticism. The result was a work of literary accomplishment new for its day. Where others had succumbed to the overwhelming scale of the subject, Dutton refused to capitulate. And more, his book was illustrated by the work of two extraordinary artists who expressed extremes of interpretive talent: Thomas Moran, an Englishman who had joined the Powell Survey at the same time as Dutton, and William Henry Holmes, who had worked with the Rocky Mountain Survey since 1872.

Moran was one of three brothers, all painters, who were much influenced by

Highsiding through Lava Falls

Lava Falls is considered the most formidable and unforgiving rapid on the Colorado. Here a supply raft for a party of dories running the river plunges through Lava's deepest "hole" and heads for Dead Man's Rock. The hole has a standing wave rising 20 feet, and the rock—in direct line for all who take the often unavoidable right-hand course—is hard, sharp, and immovable.
45. E-3.

J.M.W. Turner, the great English landscape artist of the early nineteenth century. Turner was a romantic painter and a pioneer in impressionistic use of light and color, qualities clearly reflected in Moran's work. Moran saw Grand Canyon as a place of reflected light, of contrasting color and shadow, of changing mood and expression. Though he compromised with the convention of his day by including far more detail than Turner would have allowed, his canvases are broad sweeping vistas of an irresistible, wild landscape. They capture the Canyon's mood, its size, convey the artist's excitement, and compel the response that here is one of the wonders of the Earth that one must see for oneself. With his paintings of Grand Canyon (and Yellowstone), Moran probably did more than anyone to excite the public about the American Southwest.

It is extraordinary that Moran was selected by scientists to convey the scenery. Moran made no pretense that he even tried to convey geological detail accurately, and Powell and Dutton must have known it. But what a stroke of genius it was to capture the spirit of the canyonlands through one artist and convey the strict and necessary detail of them through another. The fact that W.H. Holmes is more generally remembered in the scientific world as an anthropologist, for he became curator of anthropology at the Smithsonian Institution in 1897 and followed Powell as the chief of the Bureau of American Ethnology in 1902, is an acknowledgment of his intellectual capacity, not a reflection on his abilities as a scientific artist. In fact, his artistic merits later benefited

the National Gallery of Art, of which he was director for ten years from 1922 to 1932.

Holmes drew panoramas of plateau landscapes in meticulous detail. The problems that he solved were those of conveying scale and perspective while maintaining the integrity of his subject, another way of stating that his drawings had to be three-dimensionally acceptable to the eye although semicircular in concept. It so happens that my interests in photography have made me face, and try to solve, exactly the same problems in exactly the same localities for almost the same reasons. When I started panoramic photography in the Grand Canyon region I knew of Holmes' work and had very much admired it, but I was familiar only with his masterpiece, the view from Point Sublime on the North Rim of Grand Canyon. It was only when the U.S. Geological Survey at Reston, Virginia, generously produced portfolios of Holmes' drawings from their archives and allowed me to browse through them that, as I carefully turned the precious pages of his field notebooks, I discovered to my growing astonishment that for years I had innocently and literally been following in the great man's footsteps. Many of the panoramas I had taken were within a few hundred yards of Holmes' carefully selected positions.

Holmes must have hunted far and wide to find suitable locations for his panoramas. Some were obvious, "The Temples of the Virgen" from Dutton's book, for example. This was drawn from the mesa-like tiers rising above and to the south of the small town of Rockville, Utah, with Mount Kinesava dominating the central scene to

Glen Canyon Dam, Page, Arizona (overleaf)

Completed in 1966, this dam is an epitaph to one of the most beautiful and idyllic canyons on the Colorado. Glen Canyon is now buried under five hundred feet of water in Lake Powell, which stretches a hundred miles back from the dam. Conservationists deplore its presence on the grounds that it has upset the local ecology, has despoiled natural wilderness beauty, and is having an adverse effect on Grand Canyon. On the other hand, hundreds of thousands of visitors to Lake Powell can explore the canyonlands back country, and many Southwestern towns benefit from the electricity produced by the dam. Meanwhile, a coal-burning plant built by Navajo Indians on their nearby reservation belches pollution into the clear desert air and onto some of America's finest scenery. Whatever the pros and cons of the situation created by the dam, the dam itself has the distinction of being almost invisible from any distance over half a mile. 46. D-5.

the north, the West Temple and Zion Canyon to the right of it, and the rugged hills near the Cougar Mountain Fault and Crater Hill in the half distance to the left. Holmes first drew the scene as a straightforward panorama as viewed from the nearest good vantage point; then he moved back a considerable distance to sketch Smithsonian Butte to his left and Eagle Crag to his right. These additional drawings were later added to the main panorama. In my panorama of the same scene, unaware of Holmes' work in the area, I wanted to capture a moment that would characterize Zion as I like to remember it—a particular light and mood, as well as a broad hint of the geology. About one third of the total length of that shot, taken on an autumn evening after heavy rain, is reproduced on pages 30–31.

I suspect that Holmes' "Temples of the Virgen" was one of his early panoramic drawings, and that he subsequently found that the most desirable effect was obtained by looking down upon a scene rather than being on a level with it. One of the best examples of this technique is the picture Holmes drew from the top of Vulcan's Throne above Toroweap looking upriver over the Esplanade. The final version of this drawing includes details of cliff faces of the most delicate intricacy and accuracy. The foreground is, however, entirely imaginary and the men and horses that Holmes added to this part of the scene exaggerated its scale. Perhaps I should add that the foreground from Vulcan's Throne toward the Esplanade is completely uninspiring and that Holmes usually made such cosmetic adjustments to his masterpieces.

During my overnight stop on the northern edge of the Green Desert, the Jeep hooked to the rear of the motor home had been coated with a thick layer of frozen vapor from the generator exhaust. The Jeep sparkled like a Christmas cake decoration in the early morning light as I started across the hundred miles or so of snow-flurried desert road, which stretched before me. The sage and rabbit brush bristled with hoar frost on the roadside. Slowly the white-shrouded Henry Mountains began to rise above the far horizon, their summits peering remotely above an encircling halo of still cloud set against an azure sky. The canyon trinity lay beyond them. The winter photography planned for Bryce Canyon and thwarted so often by impossible weather might just come off this time. I thought of Holmes and the hardships he had faced as a young man as he sketched in frigid cold or burning sun. No motor home, no Jeep, perhaps a mule to carry his equipment and stores, and a horse to ride. I imagined Powell, Dutton, a host of other geologists, and wondered at the millions of people from all over the Earth, who had followed them.

I remembered a comment that Dutton had made in his *Tertiary History*, something about "an innovation in modern scenery." But how did the passage go on? When I reached Bryce Canyon I turned to the National Park's copy of the book and recalled his words. "As with all great innovations," Dutton had written referring to Grand Canyon, "it is not to be comprehended in a day or a week, nor even in a month. It must be dwelt upon and studied, and the study must comprise the slow acquisition of the meaning and spirit of that marvelous scenery which characterizes the Plateau Country, and of which the Great Chasm is the superlative manifestation . . ." Next morning dawned bright and clear. As I trudged down the Navajo Loop on snowshoes with camera pack on my back and ski poles to steady me, my heart sang. I was back in the Plateau Country and my luck was in.

ACKNOWLEDGMENTS

The courtesy, helpfulness, and general knowledge of National Park Service rangers are some of the memorable impressions of a locality retained by visitors. When these visitors are foreigners like myself, although somehow one never feels a "foreigner" in the United States, that impression can sometimes be enhanced by contrasting experiences elsewhere. If, as a serious student of the natural features of a particular park, you were to work behind the scenes at park headquarters, as I have been privileged to do, you would find yourself in an unmatched world of information, firm control, and cordiality, in which anything within reason is possible, and anything which is not is subject to approval by the superintendent. Some of my requests fell into the latter category. All were patiently reviewed and none were refused.

Much of the photography in this book simply could not have been attempted without the interest and support of park superintendents and their staff. My grateful thanks are therefore most warmly given to the following personnel in particular and to many other tremendously helpful Park Service people responsible to them. *Grand Canyon National Park:* Superintendent Merle Stitt and Management Assistant Roger Giddings; Chief Park Ranger Gary Kuiper and Assistant C.P.R. Dick McLaren; Librarian Louise Hinchliffe. *Zion National Park:* Superintendent Bob Heyder (now of Mesa Verde) and Chief Park Naturalist Vic Jackson. *Bryce Canyon National Park:* Superintendent Tom Hobbs, Chief of Interpretations & Resource Management Tom Henry, and Naturalist John Bezy. *Petrified Forest National Park:* Superintendent David Ames and Chief Park Ranger Andy Ringgold. *Wupatki National Monument:* Superintendent Marjorie Hackett and Chief of Interpretations Bill Palek.

Martin Litton, master boatman of the Colorado River, is a delightful personality and a very generous man. A special word of thanks to Martin and the men of Grand Canyon Dories whom he led on a particularly adventurous and enjoyable Grand Canyon traverse on my behalf. Powell would have given his all to have had the skills of boatmen Kenton Grua, "Bego" Gerhart, "Fleet" Eakland, Tim Cooper, and "Wren" Reynolds at his disposal, and the cooking ingenuity of Kenly Wiles and Ellen Tibbetts (a no mean boatperson herself). And a word of thanks to Joel Mensik to whom a few of us owe the exhilarating experience of running Specter Rapid backwards and one or two other exciting excursions in the "Soaky Steps."

A word of appreciation to reconnaissance pilot Lynn Page and for the delicate touch of helicopter pilots Dan O'Connell and Randy Stewart of Grand Canyon, and Jim Clark of Cedar City, Utah, who thankfully never failed to remember where they had dropped me off.

My special thanks to Tom Gamble, Chief of Glen Canyon Field Division, U.S. Bureau of Reclamation, for showing me the other side of the coin; to Jim Balsley, Assistant Director of Land Resources U.S.G.S., Reston, for making Holmes' field notebooks available from the Survey's archives and many other favors; and to Peter Adams, Deputy Curator, Geological Museum, London, who first persuaded me in this direction, and for advice with respect to technical illustrations. And to Dr. Edwin D. McKee (U.S.G.S.), Dr. Richard G. Beidleman (Colorado College), and Dr. Robert C. Euler (Grand Canyon N.P. Anthropologist) for reading individual chapters, saving me from many errors and warning me severely of the gray areas I had entered. If errors remain they are entirely my own.

A large number of very skilled people are intimately involved in the production of a book of this character. They have surmounted technical problems of considerable dimension and their efforts are not only gratefully acknowledged but deeply appreciated. But there are three men who have worked with me from concept to conclusion, to whom I would like to pay a special tribute. First and foremost my editor, Roger Jellinek of Times Books, who has not only steered me through the rough waters of syntax and lucidity, but has persistently encouraged, occasionally extolled, and always inspired. Then my good friend Irwin Glusker of New York, book designer extraordinary—a man of wit, charm, drive, and vast experience, a delight to work with. And Gary Hincks of Diss, Norfolk, England, certainly one of the finest young technical illustrators of today, who has patiently overcome the apparently insurmountable.

Lastly, my thanks to F.W. Dunning, Curator, Geological Museum, London, for his thoughtful reading of the text for the Reader's Digest Edition. His comments and suggestions were of great value.

To you all, thank you.

BIBLIOGRAPHY

Anthropology

Anderson, Douglas D. "A Stone Age Campsite at the Gateway to America." Paper dated 1968. Reprinted in *Early Man in America*. San Francisco: W.H. Freeman and Co., 1973.

Bahti, Tom. *Southwestern Indian Ceremonials*. Las Vegas, Nev.: KC Publications, 1970.

Euler, Robert C. "John Wesley Powell and the Anthropology of the Canyon Country," *Geological Survey*. Washington, D.C.: United States Government Printing Office, 1969.

Haag, William E. "The Bering Strait Land Bridge." Paper dated 1962. Reprinted in *Early Man in America*. San Francisco: W.H. Freeman and Co., 1973.

Holloway, Ralph L. "The Casts of Fossil Hominid Brains." Paper dated 1974. Reprinted in *Human Ancestors*. San Francisco: W.H. Freeman and Co., 1979.

Hughes, J. Donald. *In the House of Stone and Light: A Human History of the Grand Canyon*. Flagstaff, Ariz.: Grand Canyon Natural History Association, 1978.

Jennings, Jesse D. *Ancient Native Americans*. San Francisco: W.H . Freeman and Co., 1978.

Kluckholn, Clyde, and Dorothea Leighton. *The Navaho*. Garden City, N.Y.: Doubleday, 1962.

Leakey, Richard E., and Roger Lewin. *Origins*. New York: E. P. Dutton, 1977.

Martineau, LaVan. *The Rocks Begin to Speak*. Las Vegas, Nev.: KC Publications, 1973.

McGregor, John C. *Southwestern Archaeology,* 2nd ed. Urbana, Chicago, London: University of Illinois Press, 1965.

Museum of Northern Arizona. *An Introduction to Hopi Kachinas*. Flagstaff, Ariz.: The Arizona Bank, 1977.

Powell, Major J.W. *The Hopi Villages: The Ancient Province of Tusayan*. Palmer Lake, Colo.: Filter Press, 1972.

Schroeder, Albert H. *Of Men and Volcanoes: The Sinagua of Northern Arizona*. Globe, Ariz.: Southwest Parks and Monuments Assoc., 1977.

Schwartz, Douglas W. "A Historical Analysis and Synthesis of Grand Canyon Archaeology." *American Antiquity*, Vol. 31, No. 4 (April 1966), p. 469.

———. "Nankoweap to Unkar: An Archaeological Survey of the Upper Grand Canyon." *American Antiquity*, Vol. 30, No. 3 (March 1965), p. 278.

———, Michael P. Marshall, and Jane Kepp. *Archaeology of the Grand Canyon: The Bright Angel Site*. Santa Fe, N. Mex.: School of American Research Press, 1979.

Siegrist, Roland, ed. *Prehistoric Petroglyphs and Pictographs in Utah*. Salt Lake City, Utah: Utah State Historical Society, 1972.

Simons, Elwyn L. "The Early Relatives of Man." Paper dated 1964. Reprinted in *Human Ancestors*. San Francisco: W.H. Freeman and Co., 1979.

Solecki, Ralph. "How Man Came to North America." Paper dated 1951. Reprinted in *Early Man in America*. San Francisco: W.H. Freeman and Co., 1973.

United States Department of the Interior Geological Survey. *Tree Rings: Timekeepers of the Past*. Washington, D.C.: U.S. Department of the Interior, 1978.

Wormington, H.M. *Prehistoric Indians of the Southwest*. Denver: The Denver Museum of Natural History, 1971.

Geology

Breed, William J. *Investigations in the Triassic Chinle Formation*. Museum of Northern Arizona Bulletin #47. Flagstaff, Ariz.: Museum of Northern Arizona, 1972.

Butzer, Karl W. *Geomorphology from the Earth*. New

York: Harper & Row, 1976.
Cameron, A.G.W. "The Origin and Evolution of the Solar System." Paper dated 1975. Reprinted in *The Solar System*. San Francisco: W.H. Freeman and Co., 1975.
Colbert, Edwin H. *Geology of the Grand Canyon*. Flagstaff, Ariz.: Museum of Northern Arizona and Grand Canyon Natural History Assoc., 1976.
Dewey, John F. "Plate Tectonics." Paper dated 1972. Reprinted in *Continents Adrift and Continents Aground*. San Francisco: W.H. Freeman and Co., 1976.
Dietz, Robert S. "Geosynclines, Mountains, and Continent-Building." Paper dated 1972. Reprinted in *Continents Adrift and Continents Aground*. San Francisco: W.H. Freeman and Co., 1976.
———, and John C. Holden, "The Breakup of Pangaea." Paper dated 1970. Reprinted in *Continents Adrift and Continents Aground*. San Francisco: W.H. Freeman and Co., 1976.
Eicher, Don L. *Geologic Time*, 2nd ed. Englewood Cliffs, N.J.: Prentice-Hall, 1968.
Fenton, Carroll Lane, and Mildred Adams Fenton. *The Fossil Book: A Record of Prehistoric Life*. Garden City, N.Y.: Doubleday, 1958.
Ford, Trevor D., and William J. Breed. *Geology of the Grand Canyon*. Flagstaff, Ariz.: Museum of Northern Arizona and Grand Canyon Natural History Assoc., 1976.
Frakes, L.A. *Climates Throughout Geologic Time*. New York: Elsevier Scientific Publishing Co., 1979.
Gilluly, James. *Principles of Geology*, 4th ed. San Francisco: W.H. Freeman and Co., 1951.
Hallam, A. "Alfred Wegener and the Hypothesis of Continental Drift." Paper dated 1975. Reprinted in *Continents Adrift and Continents Aground*. San Francisco: W.H. Freeman and Co., 1976.
———. "Continental Drift and the Fossil Record." Paper dated 1972. Reprinted in *Continents Adrift and Continents Aground*. San Francisco: W.H. Freeman and Co., 1976.
Hamblin, W. Kenneth. "Late Cenozoic Volcanoism in the Western Grand Canyon," *Geology of the Grand Canyon*. Flagstaff, Ariz.: Museum of Northern Arizona and Grand Canyon Natural History Assoc., 1976.
———, and J. Keith Ridby. *Guidebook to the Colorado River, Part 1: Lee's Ferry to Phantom Ranch in Grand Canyon National Park*. Brigham Young University Geology Studies, Vol 15, Part 5. Provo, Utah: Brigham Young University, 1968.
———. *Guidebook to the Colorado River, Part 2: Phantom Ranch in Grand Canyon National Park to Lake Mead, Arizona-Nevada*. Brigham Young University Geology Studies, Vol. 16, Part 2. Provo, Utah: Brigham Young University, 1969.
Heirtzler, J.R. "Sea-Floor Spreading." Paper dated 1968. Reprinted in *Continents Adrift and Continents Aground*. San Francisco: W.H. Freeman and Co., 1976.
Hunt, Charles B. *Geology of the Grand Canyon*. Flagstaff, Ariz.: Museum of Northern Arizona and Grand Canyon Natural History Assoc., 1976
———. *Natural Regions of the United States and Canada*. San Francisco: W.H. Freeman and Co., 1968.
Kurten, Björn. "Continental Drift and Evolution." Paper dated 1969. Reprinted in *Continents Adrift and Continents Aground*. San Francisco: W.H. Freeman and Co., 1976.
McKee, Edwin D. "Paleozoic Rocks of Grand Canyon." *Geology of the Grand Canyon*. Flagstaff, Ariz.: Museum of Northern Arizona and Grand Canyon Natural History Assoc., 1976.
———. *Ancient Landscapes of the Grand Canyon Region: The Geology of Grand Canyon, Zion, Bryce, Petrified Forest and Painted Desert*, 26th ed. Flagstaff, Ariz.: Northland Press, 1931.
———. *The Environment and History of the Toroweap and Kaibab Formations of Northern Arizona and Southern Utah*. Washington, D.C.: Carnegie Institution of Washington, 1938.
———. ed. *Evolution of the Colorado River in Arizona: An Hypothesis Developed at the Symposium on Cenozoic Geology of the Colorado Plateau in Arizona, August 1964*. Flagstaff, Ariz.: Museum of Northern Arizona, 1967.
Pewe, T.L., and R.G. Updike. *San Francisco Peaks: A Guidebook to the Geology*, 2nd ed. Flagstaff, Ariz.: Museum of Northern Arizona, 1976.
Press, Frank, and Raymond Siever. *Earth*, 2nd ed. San Francisco: W.H. Freeman and Co., 1974.
Shelton, John S. *Geology Illustrated*. San Francisco: W.H. Freeman and Co., 1966.
Siever, Raymond. "The Earth." Paper dated 1975. Reprinted in *The Solar System*. San Francisco: W.H. Freeman and Co., 1975.
Steel, Rodney, and Anthony P. Harvey, eds. *The Encyclopedia of Prehistoric Life*. New York: McGraw-Hill Book Co., 1979.
Sullivan, Walter. *Continents in Motion: The New Earth Debate*. New York: McGraw-Hill Book Co., 1974.
Toksöz, M. Nafi. "The Subduction of the Lithosphere." Paper dated 1975. Reprinted in *Continents Adrift and Continents Aground*. San Francisco: W.H. Freeman and Co., 1976.
United States Department of the Interior. *The Antarctic and Its Geology*. Washington, D.C., 1975.
———. *Earthquakes*. Washington, D.C., 1974.
———. *Geologic Time*. Washington, D.C., 1978.
———. *The Interior of the Earth*. Washington, D.C., 1976.
———. *Our Changing Continent*. Washington, D.C., 1978.
Wilson, J. Tuzo. "Continental Drift." Paper dated 1963. Reprinted in *Continents Adrift and Continents Aground*. San Francisco: W.H. Freeman and Co., 1976.

Paleobiology

Barghoorn, Elso S. "The Oldest Fossils." Paper dated 1971. Reprinted in *Life: Origin and Evolution*. San Francisco: W.H. Freeman and Co., 1979.
Colbert, Edwin H. "The Ancestors of Mammals." Paper dated 1949. Reprinted in *Evolution and the Fossil Record*. San Francisco: W.H. Freeman and Co., 1978.
Deering, R.A. "Ultraviolet Radiation and Nucleic Acid." Paper dated 1962. Reprinted in *Conditions for Life*. San Francisco: W.H. Freeman and Co., 1976.
Dickerson, R.E. "Chemical Evolution and the Origin of Life." Paper dated 1978. Reprinted in *Evolution*. San Francisco: W.H. Freeman and Co., 1978.
Dobzhansky, Theodosius, et al. *Evolution*. San Francisco: W.H. Freeman and Co., 1978.
Folsome, Edwin Clair. *The Origin of Life*. San Francisco: W.H. Freeman and Co., 1979.
Fox, C. Fred. "The Structure of Cell Membranes." Paper dated 1972. Reprinted in *Conditions for Life*. San Francisco: W.H. Freeman and Co., 1976.
Frieden, Earl. "The Chemical Elements of Life." Paper dated 1972. Reprinted in *Conditions for Life*. San Francisco: W.H. Freeman and Co., 1976.
Glaessner, Martin F. "Pre-Cambrian Animals." Paper dated 1961. Reprinted in *Evolution and the Fossil Record*. San Francisco: W.H. Freeman and Co., 1978.
Howells, William W. "The Distribution of Man." Paper dated 1960. Reprinted in *Evolution and the Fossil Record*. San Francisco: W.H. Freeman and Co., 1978.
Lewis, John S. "The Chemistry of the Solar System." Paper dated 1972. Reprinted in *Life: Origin and Evolution*. San Francisco: W.H. Freeman and Co., 1979.
Margulis , Lynn. "Symbiosis and Evolution." Paper dated 1971. Reprinted in *Life: Origin and Evolution*. San Francisco: W.H. Freeman and Co., 1979.
May, Robert M. "The Evolution of Ecological Systems." Paper dated 1978. Reprinted in *Evolution*. San Francisco: W.H. Freeman and Co., 1978.
Newell, Norman D. "The Evolution of Reefs." Paper dated 1972. Reprinted in *Evolution and the Fossil Record*. San Francisco: W.H. Freeman and Co., 1978.
Runcorn, S.K. "Corals as Paleontological Clocks." Paper dated 1966. Reprinted in *Evolution and the Fossil Record*. San Francisco: W.H. Freeman and Co., 1978.
Schopf, J. William. "The Evolution of the Earliest Cells." Paper dated 1978. Reprinted in *Evolution*. San Francisco: W.H. Freeman and Co., 1978.
Seilacher, Adolf. "Fossil Behavior." Paper dated 1967. Reprinted in *Evolution and the Fossil Record*. San Francisco: W.H. Freeman and Co., 1978.
Valentine, James W. "The Evolution of Multicellular Plants and Animals." Paper dated 1978. Reprinted in *Evolution*. San Francisco: W.H. Freeman and Co., 1978.
Wald, George. "Life and Light." Paper dated 1959. Reprinted in *Conditions for Life*. San Francisco: W.H. Freeman and Co., 1976.
———. "The Origin of Life." Paper dated 1954. Reprinted in *Life: Origin and Evolution*. San Francisco: W.H. Freeman and Co., 1979.
Washburn, Sherwood L. "The Evolution of Man." Paper dated 1978. Reprinted in *Evolution*. San Francisco: W.H. Freeman and Co., 1978.

Zoology

Dodge, Natt H. *Poisonous Dwellers of the Desert*. Globe, Ariz.: The Southwestern Monuments Assoc., 1951.
Hoffmeister, Donald F. *Mammals of Grand Canyon*. Urbana, Ill.: University of Illinois Press, 1971.
Larson, Peggy with Lane Larson. *The Deserts of the Southwest*. San Francisco: Sierra Club Books, 1977.
Moenke, Helen. *Ecology of Colorado Mountains to Arizona Deserts*. Denver: Denver Museum of Natural History, 1971.
Nelson, Ruth Ashton. *Plants of Zion National Park: Wildflowers, Trees, and Shrubs*. Springdale, Utah: Zion Natural History Assoc., 1976.
Palmer, Laurence E. *Fieldbook of Natural History*, 2nd ed. New York: McGraw-Hill Book Co., 1949.
Wauer, Roland H. *Reptiles and Amphibians of Zion National Park*. Springdale, Utah: Zion Natural History Assoc., 1964.
———, and Dennis L. Carter. *Birds of Zion National Park and Vicinity*. Springdale, Utah: Zion Natural History Assoc., 1965.
Welsh, Stanley L. *Flowers of the Canyon Country*. Provo, Utah: Brigham Young University Press, 1971.

INDEX

Page numbers in italics refer to illustration captions.

A

Africa: camel, evolution of, 145; continental drift, 13, 21, 139; *Glossoptera* fossils in, 12; hominid fossils in, 152; human evolution, 151, 153
Air pollution, *191*
Alaskan Refuge, 154
Alligator, sunset view from, *185*
Anasazi Indians, 155, *157*, 159, 162
Ancestral Colorado River, *13*, 40, 41, 69
Animal life, 95; adaptive radiation, and convergence, 145–47, *146–47*, 148; amphibians, 91, 95, 140, 142, 143; annelids, 86, 88, 131; arthropods, 87–88, *120*, 135–36; birds, 142; camel, 145; corals, 134–35, 143; of Cretaceous Period, 142, 144–45, *144–45*; of Devonian Period, 140; dinosaurs, 25, 28, 95, 142–43, 144; dispersion, 145–47; echinoderms, 90; exoskeletons, 134; fish, 28, 91, 140, 143; fossils, *see* Fossils; land forms, development of, 135–36; mammals, 144–45, *144–45*; mammals, Nuttall's test for genetic relationships, 151–52; mass extinctions, 95, *120*, *121*, 143–44; reproduction and survival, *140*, 140–42; reptiles, 95, 140–43, 144–45, *144–45*; of Triassic Period, 140–41; trilobites, 87–90, 91, 95, *120*, 131, 143; vertebrates, 139–43
Antarctica, 11–13, *23*, 24–25
Apache Indians, 179
Aquarious Plateau, 44
Araucarioxylan, *133*, 138, 139
Archeozoic Era, 27
Artifacts of prehistoric man, 153; dating, dendrochronology and, 155; Desert Culture, 151, 154; Hopi, 159
Asia, migration of man from, 153–54
Astronauts, geology training of, 51–53
Australia, 139
Avella, Pennsylvania, human site in, 153

B

Bacon, Francis, 13
Baja California, and continental drift, 21
Banded Iron Formation, 56
Barghoorn, Elso S., *119*
Barrett, Peter J., 24
Basin Mountains, 32, 33, *33*
Bass Limestone, 111, 123
Bass Rapid, 189
Bass, William Wallace, 189
Battleship Butte, *73*
Beardmore Glacier, Antarctica, 11–13, 24–25
Bering Strait, 153–54
Beringia, 154
Bernal, J. D., 118–19
Billingsley, George H., 78
Biochemistry, and evolution of life, 114–20
Biosphere, 122–23, 144
Biostratigraphy, 29, 85–91
Boucher Rapid, *147*
Bowers, Henry R., 12
Boysag Point, *89*; view from, *159*
Bright Angel Canyon, *38*, 67, *73*, 75
Bright Angel Creek, Indian settlement along, 158
Bright Angel Shale, 87
Bruno, Giordano, 83, 84
Bryce Amphitheater, *171*
Bryce Canyon, 25, 31, *36–37*, 44; formation, 44; fossils, 87, 136; life forms, 145, 147; photographing, *171*; rocks of, 44, 48–49, 57, *62*, 123; unconformities, 65–67, *66*, 78–79, 136; vulcanism, 99, *100;* youngest formation at, 48
Bryce Canyon National Park, 189

C

Cabeza de Vaca, Álvar Núñez, 178
Cambrian Period, 28; day and year lengths, 135; fossils, 87, 88, 90; life forms in, 134; rocks of, 67–68, 77; trilobites, *120*
Cárdenas, García López de, 178–79
Carmel Limestone, 48
Carson, Col. Kit, 179
Cedar Breaks, *36, 137*
Cedar Butte, 102
Cenozoic Era, 27, 28; climates of, 123; formations, disappearance of, during, 44; life forms in, 144–47; rocks of, *62*
Cheops Pyramid, *73, 83*
Chinle Formation, *51, 57, 75*, 139, 142
Christianity, and geology, 83–84
Chuar Butte, view from, *116–17*
Chuar Formation, eukaryotes in, 123–31
Cibola, Spanish search for, 178–79
Climates, 12, 13; Antarctic, 12; arkose as indicator of, 56; effects of, *71;* structures as evidence of, 59, 123; temperatures, life and, 122–23
Coalpits Wash, 145
Coalsack Bluff, Antarctica, 24–25
Coconino Sandstone, 95
Cohonina Indians, 155, 158, 162
Colbert, Edwin H., 24–25
Colorado Plateau *(see also* specific features), *17–18;* climate, 59; elevation, 79; map of panorama points, locations and roads, *14;* peneplanation, ultimate, 79; physical features, principal *(see also* specific features), *19;* position, continental drift and, *23*, rocks *(see also* Rocks and rock formations), 25, 57; uplift of, 31–34, *33*, 40, 99, 139; vulcanism in, 98, 99
Colorado River *(see also* specific features), *39;* Ancestral, *13*, 40, 41, 69; Canyon formation, 34–41, *35, 39;* as evolutionary barrier, 145–47, *146–47;* explorations of, 186–88; flow pattern, 40; Indian settlement along, 155–58, *157;* lava dams, 106; salt collected along by Indians, 159; views of, *11, 67, 103, 146–47, 185, 187*
Conquistadores, 163, 165–79
Continental drift: coastlines, fit of, 13; convective movement of asthenosphere and, 21–24, *22;* convective movement and lithosphere, 56–57; ecology, life forms and, 134, 136–37; fossil evidence, 13, 24–25; Grand Canyon region as showcase of events of *(see also* specific features), 25–26; life forms and, 12–13, 134–37, 138–39, 145, 153; and magnetized rock, 13–20, *21;* theory, 13–25; theory, opposition to, 13
Continental plates, 21, *23;* Colorado Plateau uplift, 32–34; and vulcanism, 98
Continents *(see also* Continental drift; names), 12–13, *23*
Convective movement, 21–24, *22;* rock forms and, 56–57
Coral Pink Sand Dunes, 59, *61*
Coronado, Francisco Vasquez de, 178
Cougar Mountain Fault, 193
Court of the Patriarchs, *91*
Crater Hill, 193
Cretaceous Period, 28; Grand Canyon formation during, 34–40; life forms in, 142, 144–45, *144–45;* mass extinction during, 143; rocks missing, 79; rocks of, 49
Cuvier, Baron Georges, 85

D

Dakota Sandstone, 79
Dams: lava, *102*, 106–10, *107;* man-made, 106, 165, *187, 191*
Dana Butte, *113*
Darwin, Charles, 13, 85
Dating, geological: by coral growth bands, 135; dendrochronology, 155; fossils and evolutionary order, 27–28, *28*, 85; radiometry, 60–61; time scale of, 27–29, *28, 29*
da Vinci, Leonardo, 84, 85
Day, length of, 135
Dead Man's Rock, *190*
de Niza, Fray Marcos, 178
Desert View, *38, 83*, 178, 179
Devonian Period, 28; amphibians of, 140; corals of, as paleontological clocks, 135; day length, 135; fish evolution during, 28, 140; fossils of, 91; rocks of, 77, 90–91; trilobites in, 90
Diamond Creek area, 78, *166*
Douglass, A. E., 155
Dox Formation, 111
Dutton, Clarence Edward, *169*, 189, 193

E

Eagle Crag, 193
Earth: age of, 29, 85; atmosphere, evolution of, 53–55, *54*, 56, 120, 122–23; atmosphere, organic synthesis and, 114–18; biosphere, 122–23, 144; continents *(see also* names), 13–25, *23;* evolution, time scale of, *28*, 29, *29;* formation of, 20–21, *22*, 51, 52; geological eras *(see also* names), 25–29, *28, 29;* isotasy, 33, 189; "life" on *(see also* Evolution; Life and life forms), *54*, 114–31; ozone layer, 120; polarity, 13–20, *20;* rhythms of, and biorhythms, 133; rotation and length of day, 135; temperatures, 122–23
Earthquakes, 99, 100, 110, 139
Ecosystems, 134
Egloffstein, Baron F. W. von, *166*, 188
Elden Mountain, 102
Elliot, David, 24
Entrada Sandstone, 79
Erosion *(see also* specific structures): chemical weathering, 41, 69; by glaciation, 68–69; meteorite, 53–54; monadnocks resistant to, 69, *77;* Moon, 52; peneplanation, 67–69, 75; permeability to water, 45, *45;* by rivers, 34, *35*, 41; of spire shapes, 48–49; and unconformities *(see also* Unconformities), 65–67, *66*
Escalante Butte, 42–44; views from, *11, 13*
Escarpments, 44
Esplanade, 94, 95, 106, 193; views of, *159, 169*
Esteban, 178
Eurasia, continental drift and dispersion of life forms, 145, 153
Europe, and continental drift, 13, 139
Evans, Edward R., 12
Evolution *(see also* Animal life; Life and life forms; Plant life); algae, *119*, 120, 131, 133, 134; arthropods, 135–36; atmosphere and, *54*, 120, 122–23; bacteria, 119–20, 131; birds, 142; and dating, geological, 85; DNA, 115, 118, 123; ecological replacement, *121;* eukaryotes, 119, *119*, 120, 123–31, 137; extinction, periods of, 144; fish, 28, 140; fungi, 131, 133; genetic relationships, Nuttall's immunological test for, 151–52; geological eras and *(see also* names), 28–29, *28;* Grand Canyon as evolutionary barrier, 145–47, *146–47, 148;* humanoid, *28*, 29, 151–53; to "living" cell, 119; mutation, 118; natural selection, 123, 144; organic, *54*, 114–19, *118;* organic, laboratory synthesis of, 114–15; prokaryotes, 119–20, 123, 131; reproduction processes, 123, 131, (of animals), 140–42, (of plants), 138; sea to land, transition from, 134–43; species, number of, 85; symbiosis or mutualism, 131, 133; trilobites, extinction of, 120; vertebrates, 139–47
Exploration of the Colorado River and Its Canyons (Powell), 186, 187

F

Faults: Cougar Mountain, 193; Grand Canyon, *42;* Kaibab Plateau, *43;* movement, and earthquakes, 99, 110; Hurricane, 107, 110; San Andreas, 33; Toroweap, 106, 107, *107*, 110; types of, *43*, vulcanisms and, 98, 99
Fish, 143; evolution, 140; fossils, 28, 91, 140
Folsome, Clair E., 119
Fort Rock Caves, Oregon, 153
Fossils *(see also* Life and life forms): age of, 60; algae, 43, 111; animal life, 24–25, 43, 48, 85–95, *86, 87, 89*, 142, 143; annelids, 86, 131; Antarctic, 12–13, 24–25, *Araucarioxylon*, *133; Bothriolepis*, 91, 140; of Cambrian Period, 87, 88, 90; carbonization, 85; Christian theology on, 83, 84; continental drift theory, 13, 24–25; *Cooksonia*, 138; coprolites, 85; of Devonian Period, 91; dinosaur, 142; ecological replacement evidenced by, *121;* evolutionary order, geological dating and, 27–28, *28*, 85; fish, 28, 91; gastroliths, 85; geological eras *(see also* names), 27–29, *29; Glossoptera*, 12, 24; hominid, 152, 153; human, 153; index-fossils, 29, *83;* of Jurassic Period, 142; *Lystrosaurus*, 25; marine, 28, 84, *89*, 90–91, 94, 95, 111, *121;* of Mesozoic Era, 138; microfossils, *119;* of Mississippian Period, 91; oldest in Grand Canyon, 111; ostracodes, 91; of Pennsylvanian Period, 43, 78; of Permian Period, 43, 95; petrifaction, 85–86, plant life, 12–13, 78, 111, *133*, 138, 139, 144; of Pleistocene Epoch, 153; of Precambrian Era, 111; *Ramapithecus*, 153; reef ecology and, 136–37; of Silurian Period, 138; stromatolites, 111, 134, 137; trace fossils, 86, *86, 87*, 95, 140, 142; trees, petrified, *133*, 138, 139; of Triassic Period, 138, 139–40, 142; trilobites, *83*, 87–90, 91, 95, 131; and unconformities, 136; vertebrates, 139, 142, 143
Four Corners Country, 31–32; Carson's rout of Navajo, 179–86
Fox, Sydney, 118

G

Garces, Fray Francisco Tomás, 178
Geological eras *(see also* specific features, formations, names, subjects), 27–29, *28, 29*
Geology, development of, 27–29, 83–85
Glaciation, effects of, 68–69
Glen Canyon Dam, 165, *187, 191*
Gondwanaland, 12, *23,* 25, 134, 136, 139
Grand Canyon *(see also* specific features, subjects), 13; in art, *166–69, 187,* 188, 189–93; continental drift evidenced in, 25–26; distances, viewed from Hopi Point, *146–47;* as evolutionary barrier, 145–47, *146–47, 148;* exploitation of, 189; exploration of, 186–93; fault section through, *42;* formation, 25–29, *28,* 34–41, fossils, 43, 48, 87–95, 111, 136, 140; Indians in region of *(see also* Indians; names), 155–59, *157, 159, 163;* Kaibab Plateau and, 39; life zones of, *148;* man in, 151–93, name, 189; placenames, 189, rocks of *(see also* Rocks and rock formations), *13, 28,* 29, 45, 57, 61, 65, 67, 90–91, 111, 123; storm over, *65, 73;* unconformities of, 65–67, *66,* 78–79, 136; vulcanism in, 98, 99, *100,* 111
Grand Canyon Game Preserve, 189
Grand Canyon Series, 69–75, 95, 111
Grand Canyon National Park, 189
Grand Falls, Little Colorado River, *35, 102,* 106
Grand River, 187, 188
Grand Wash Cliffs, *33, 38, 42*
Granite Gorge, 26, *27,* 67, 68, 75, 111; alpine mountains of, *80;* rocks of, 57, 61
Graphite Peak, Antarctica, 24
Gray Cliffs, 44
Great Thumb Mesa ruins, 158
Great Unconformity, *66,* 67
Greece, Ancient, and uniformitarianism, 83–84, 85
Green River, 41, 187, 188
Gunflint Lake, Minnesota, microfossils of, *119*

H

Hakatai Shale, 111
Hance, John, 189
Hance Rapid, *11,* 111, 123, 189
Hance Trail, 189
Havasu Canyon, *38,* 158; Indian settlement in, 158–59, *159,* mining, 189; name derivation, 94; seen from Boysag Point, *159*
Havasu Creek, *155*
Havasu Point, view from, *80*
Havasupai Indians, 94, 158–59, *159,* 178, 179
Havasupai Point, *146*
Hawaiian Islands, vulcanism and, 98, 99
Hawikwah village, 178
Headquarter Schists, Wyoming, 68
Henry Mountains, 193
Hermit Shale, 95
Herodotus, 83, 85
Hillers, John K., 163
Holmes, William Henry, *166, 169,* 189, 190–93
Hominids, evolution of, 152–53
Homo erectus, 28, 152, 153
Homo sapiens, 152, 153
Homo sapiens neanderthalensis, 152, 153
Hopi (Hopituh) Indians, 159; Navajo raids on, 179; religious beliefs, 159–62, *163*
Hopi Point, *113;* sunset view from, *146–47, 185*
Hualapai Drainage system, 40, 41, 69
Humans, evolution of, *28,* 29, 151–53
Humphrey's Peak, 97
Hurricane Fault, 107, 110
Hutton, James, 26–27, 85
Hydrosphere, *see* Oceans

I

Ice ages, 55, 123; and Bering Strait land bridge ("Beringia"), 153–54; isostatic effect of, 34; and peneplanation, 69; and sea level, 153–54
Ice caps, polar, 137
India: continental drift, 21; fossils, 12, 25
Indian Gardens, *65, 73*
Indians *(see also* names), 151, 153–63; abandonment of Grand Canyon sites, 158; bridges, wooden pole, 158; Desert Culture, 151, 154–55; early arrivals in North America, 154–55; games, 162; of Grand Canyon region, 155–59, *157, 159, 163;* Spanish subjugation of, 163, 165–79, tribal wars, 163, 179
Inner Gorge of Grand Canyon, 98
Io, vulcanism of, 97
Isthmus Link Hypothesis, 13
Ives Expedition, *166,* 187
Ives, Lt. Joseph Christmas, 187

J

Jupiter, 97–98
Jurassic Period, 28; life forms in, 142; rocks missing, 79; rocks of, 45; sand deserts, eolian, *91*

K

Kaibab Limestone, *31,* 43, 45, 95, 100; erosion, *35;* unconformity, 79, 95, 136
Kaibab Plateau: denudation, 44; faults in, *43;* uplift of, *37, 39,* 40, 44; view of New Rim of, *83*
Kaibab Sea, fossils of, *89*
Kaiparowits Formation, 79, 136
Kanab Canyon, *38*
Kanab Creek, 107; gold, 189
Kayenta Formation, 45
Kayenta Indians, 155, 159
Kolob finger canyons, *129*
Kwagunt Butte, *116,* 123

L

Lake Bidahochi, 40
Lake Powell, *191*
Lamarck, Jean-Baptiste de, 85
Laurasia, 12, *23,* 134, 136, 139; plant life in, 139; rocks of, 45
Lava Cascades of Toroweap, *103,* 110
Lava Falls, *103,* 107, 110–11; highsiding through, *190*
Life and life forms *(see also* Animal life; Evolution; Fossils; Plant life); biorhythms, 133; biosphere and, 144; categories of, 134; continental configurations and, 12–13, 134, 136–37, 138–39, 145; ecosystems, 134; extinctions, mass, of, 95, *120, 121,* 143–44; geological eras *(see also* specific eras), 26–28, *28,* 134; laboratory synthesis of organic materials necessary for, 114–15; Grand Canyon, life zones of, *148;* lichen, 131, 133; "living" cell, evolution of, 119–31; origins, primordial, of, *54,* 114–19; requirements for, 114; symbiosis or mutualism, 131, 133; transition from sea to land, 134–43
Lignier, Octave, 138
Lipan Point, 69–75; view of, *83;* views from, *36–37,* 41–44, 75
Little Colorado River, *35,* 40, 99, 102–6; flow direction, 40–41; Indian settlement along, 158; salt collected along by Indians, 159
Little Colorado River Canyon, lava dam and, *102*
Litton, Martin, *189*
Lower Granite Gorge, 68, *166*
Lyell Butte, *83*
Lyell, Sir Charles, 85

M

Man: arrival in North America, 153–54; evolution of, *28,* 151–53; Indians *(see also* Indians; names), 151, 153–55
Manakatcha, 94, 95
Marble Canyon, *67;* inclusion in Grand Canyon National Park, 189; Indian settlement in, *157,* 158; name, *165*
Markagunt Plateau, 44
Mather Point, *175*
Matkatamiba Canyon, *155*
McKee, Dr. Edwin D., 34, 40, 51
McKee, Edwin H., 40
Meadowcroft Shelter, Avella, Pa., 153
Mendoza, Antonio de, 178
Mesa Verde Indians, 155
Mesozoic Era, 27, 28; climates of, 123, desert landscape of, 59; formations, disappearance of, during, 44; fossils of, 138; life forms in, 142
Middle Granite Gorge, 68
Miller, Stanley L., 114–15
Minerals: formation, *53,* metamorphosis of, 56; mining of, 56, 189; rock coloration and, 49
Mining: Grand Canyon, 189; iron, 56
Mississippian Period, 28; fossils of, 91; rocks of, 43, 78, 91
Moenkopi formation, *31,* 79, 100, 102, 136
Moenkopi Sandstone, Sinagua dwellings of, *151,* 162
Monadnocks, 69, *77*
Moon: biorhythms and, 133; comparison of Earth and, 51–53; tides of Earth affected by, 135
Moran, Thomas, *166,* 189–90
Mormon settlers, 189
Mount Dellenbaugh, 110
Mount Kinesava, *31, 38,* 191–93
Mount Monadnock, New Hampshire, 69
Mount Sinyala, *159*
Mount Trumbull, 110, *166*
Mountains *(see also* names), alpine, 68, 80; block-faulted, 75; formation and shape, of, 21, 68, *76–77;* peneplanation, 75; rocks associated with building of, 68
Muav Limestone, 76, 77, 136
Museum of Northern Arizona, symposium at, 34–40

N

Nankoweap Creek, Indian settlement along, *157,* 158
Nankoweap Rapids, *157*
Narrows, The, of Virgin River, 44–45
National Parks, 189
Navajo Indians: culture, 179; defeat by Carson, and "Long Walk," 179–86; plant, polluting, *191;* as raiders, 163, 179
Navajo Loop Trail, *62*
Navajo Sandstone, 45–48, 59, *59, 61*
Nazca Plate, 21–24
North America *(see also* specific places, subjects): animals, dispersion of, 145–47; continental drift, 21, 139; ice age, 34, 69; man's migration to, 151–55; subduction zone, 32–33; uplift of Colorado Plateau, 31–34, *33*
North Rim: fossils, *89;* Indians of, 155–58, inclusion in Grand Canyon National Park, 189; Kaibab squirrel and other animals, 145–46, *148,* view from, 75; views of, *11, 27, 116, 146, 185*
Nuttall, George Henry Falkiner, 151–52

O

Oates, Laurence F. G., 12
Oceanic plates, 21, 23; subduction, 21–24, 32–33; and vulcanism, 98
Oceans: biosphere role of, 121–22; composition of, 121, extent of, fossils as evidence of, 84; formation and evolution of, *54,* 121; Ice Age and sea level, 153–54; life, and marine environments, 134–38, 139–40; life, origins of, and, 120; ridge-rifts in, 20, 21, *23;* tidal friction and length of day, 135; tides, and amphibian evolution, 140; transgression and regression of, 91, 94–95, 137, 153–54
Old Crow Flats, Alaska, 153, 154
Old Oraibi village (Tusayan), 159, 163, 178–79
O'Leary's Peak, view from, *97*
O'Neill Butte, *73*
Oparin, Aleksandr Ivanovitch, 115, 118
Ordovician Period, 28; fish evolution in, 140; life forms of, 135–36; rocks missing, 76–77

P

Painted Desert *(see also* specific features), 31, *42,* 99, *102;* fossils, 87; fossils, dinosaur trace, 142; Indian settlements in, 159–62, *163;* plants of, 139; rocks of, 45; trees of, *133,* 138, 139; views of, *35, 51, 116*
Painted Desert Canyon, 25
Paiute Indians, 158
Paleoclimatology, 122–23
Paleozoic Era, 27, 28; climates of, 123; corals of, as paleontological clocks, 135; rocks of, 69
Palisades of the Desert, *67, 83, 185*
Palissy, Bernard, 83, 84
Pangea, 13, 24, 42–43, *67,* 134; breakup of, *23,* 29, 134, 136, 138–39, 145; formation of, 136; rocks, visible remnants of, *13,* 43–44, 45, 49, 69, 75, 95
Panoramas: map of locations, *14;* technique of photographing, 7–8
Paria Amphitheater, *38, 45,* 48, 49
Paria River, *38,* 44
Paunsaugunt Plateau *(see also* specific features), *38,* 44, *46,* 48
Peneplanation *(see also* specific features), 67–79, *77*
Pennsylvanian Formation, 94

Pennsylvanian Period, 28; fossils of, 43, 78, (amphibian trace), 140; rocks of, 43, 78, 91
Permian Period, 28; fossils of, 43, 95; life forms in, 137, 143; rocks of, 43, 95, trilobite extinction in, *120,* 143
Petrified Forest, Painted Desert, 138, 139
Petrified Forest National Park, 138
Phantom Ranch, 51, 52, 158
Pink Cliffs, 44
Plant life: algae, *119,* 120, 131, 133, 134, 137, 138; angiosperms, 138, 144; in Cenozoic Era, 144; chlorophytes, 138; continental drift and, 12, 129; in Cretaceous Period, 144; fossils of *(see also* Fossils), 111, *133,* 138, 139, 144; land plants, evolution of, 137–39; photosynthesis, *54,* 134, 138; reproduction in, 138; rooting, 138; stromatolites, 111, 134, 135, 137; symbiosis or mutualism, 131, 133; trees, 138, 139
Pleistocene Epoch: animals, dispersion of, in, 145; *Homo sapiens,* evidences of, 153; ice age, 34, 153–54
Point Sublime: Indian settlement of, 158; views from, *27, 169*
"Portals of the Virgen" (Holmes), *169*
Powell, Maj. John Wesley, 34, 40, 98, 106–7, *107, 165,* 186–87, 188, 189, 190
Powell Plateau, 185
Precambrian Era, 27; climates, 56, 123; day and year lengths, 135; fossils of, 131; life forms, evolution of, in, *118, 119,* 131, 134; mountain structures of, 68, 69; rocks of, *11,* 43, 56, 69, *80,* 111, *117*
President Harding Rapid, Indian pole bridge at, 158
Prospect Canyon, 107, 110–11
Proterozoic Era, 27
Pueblo Indians (Puebloeans), 155; abandonment of locations, 158, 162–63; culture, 163; in Painted Desert with Hopi, 159, 163; revolt against Spanish, 179; and Spanish search for Cibola, 178–79

Q

Quarternary Garden, 28
Queen's Garden, *62*

R

Ramapithecus, 152–53
Range Mountains, 32, 33, *33*
Red Butte, 102
Redwall Cavern, *165*
Redwall Formation, 42, 43, 78, 91, 95; erosion of, 94; fossils in, 91; storm above, 65; unconformity in, 91–94
Redwall Limestone, 77, 90, 91, 136; cavern in, *165;* Indian artifacts in, 151; metamorphosis into marble, *100,* 102; uplift followed by sinkage below sea level, 78
Rims of Grand Canyon *(see also* North Rim; South Rim), *31,* 44
Rivers, erosion by *(see also* names, specific features), 34, *35,* 41
Rocks and rock formations *(see also* specific features): age, radiometric measurement of, 60–61; age, strata position and, 84; angular unconformity in, 67; of Cambrian Period, 67–68, 77; of Cenozoic Era, *62;* as climate evidence, 56, 59, 123; colors of, mineral content and, 49; of Cretaceous Period, 49, 79; of Devonian Period, 77, 90–91; erosion of, *see* Erosion; faults, types of *(see also* Faults), *43,* formation and evolution of, 52–53, *52–53,* 55–59, 65; of Jurassic Period, 45, 79, *91;* lithosphere, 21, *22;* magnetized, and continental drift theory, 13–20, *21;* minerals in, 49, *53;* of Mississippian Period, 43, 78, 91; Moon, 51, 52; nomenclature, 28–29, 94; oldest, *28,* 29; of Ordovican Period, 76–77; of Paleozoic Era, 69; peneplanation of, 67; of Pennsylvanian Period, 43, 78, 91; of Permian Period, 43, 95; of Precambrian Era, *11,* 43, 56, 69, *80,* 111, *117;* river erosion of, 34, *35,* 41; sandstone, water permeability of, 45, *45;* of Silurian Period, 76–77; of Tertiary Period, *62;* of Triassic Period, 45, 51; unconformities in, 65–67, *66,* 84; volcanoes and *(see also* Vulcanism and volcanoes), *100;* youngest, 44
Rockville, 45
Rockville Vista, *31,* 45, 49
Rocky Mountains, *33*
Ross Ice Shelf, Antarctica, 11–13

S

San Andreas fault line, 33
Sand dunes, eolian, *53,* 57–58, *58,* 59, *91,* 95
San Francisco Peaks/Mountain, *36,* 99–100; formation, 102; Hopi religious beliefs and, 162; laccoliths, 102; Sinagua Indian settlements, 162; views of, *97, 102;* volcanic eruptions, recent, 100, 162; volcanic field of, *97,* 100–6, *100*
San Juan River, 41
Schopf, James M., 24
Scott, Robert Palcon, 11–12, 165
Scripps Institute, 20
Sentinel, *91*
Sevier River, 48
Shinarump Formation, 100, 102
Shinumo Quartzite, 111
Shivwits Plateau, 100, *100*
Shoshone Point, view from, *83*
Silurian Period, 28; fossils of, 138; rocks missing, 76–77
Sinagua Indians, *151,* 162
Smith, William ("Strata"), 27, 85
Smithsonian Butte, 193
Solar energy, life and, 122
Solar system, formation of, 20
South Africa, fossils of, 12, 25
South America, 139; animals, dispersion of, 145; coastline of, 13; subduction of Nazca plate, 21–24
South American Plate, 21–24
South Rim, 25, 99; Abert squirrel, and other animals, 145–46, *148;* exploitation of, 189; Indians of, 155–58; Spaniards reach, 178; views from, 69, 75, *175, 185;* views of, *11, 116, 146*
Space exploration, 97; astronauts, geological training of, 51–53
Spanish conquest of Southwest, 163, 165–79
Steno, Nicolaus, 84
Stratigraphy, development of *(see also* Fossils; Rocks and rock formations), 26–27, 84–85
Suess, Eduard, 12
Sunset Crater, *97,* 102, 162
Supai Group, 42, 43, 78, 94–95; amphibian trace fossils, 140; Havasupai Indians of, 158–59; name derivations, 94; unconformity of, 94

T

Table Cliffs, *38, 45*
Tanner, Seth B., 189
Tanner Trail, 189
Tapeats Sandstone, 67, 69, 75
Temple Butte, 43, 76, 77
Temple Butte Limestone, 136
"Temples of the Virgen," 191–93
"Tertiary History of the Grand Canyon District" (Dutton), *169,* 189–93
Tertiary Period, 28; rocks of, *62;* vulcanism, San Francisco field, during, 100–2
Tethys Sea, 138–39
Three Patriarchs, *91*
Tonto Group, 95
Tonto Platform, 43, *73,* 75, 87
Toroweap Fault, 106, 107, *107,* 110
Toroweap Formation, 95, *103,* 169; spring view, *140;* and vulcanism, 106–11, *107*
Toroweap region, *146*
Toroweap Valley, 107–10, *166*
Tovar, Pedro de, 178
Transept, The, *167*
Tree(s), *137,* 138, 139; dendrochronology, 155; petrified, *133,* 138, 139
Triassic Period, 28; dinosaurs in, 142; fossils, 138, 139–40, 142; rocks of, 45, *51,* 139; trees of, 138, 139
Tusayan (Old Oraibi), 178–79
Tusayan on South Rim, 155–58

U

Uinkaret Plateau, *100,* 110
Unconformities *(see also* specific features), 65–67, *66,* 75–79, 84, 91; life forms, prevalence of, and, 136–37; and mass extinctions, 143
Uniformitarianism, 84, 85
Unkar Creek, 67, 69; Indian settlement along, 158; region, 49, 75, 111
Unkar Rapid, 43, *67,* 69
Upper Granite Gorge, 68, 75; view of, *11*
Upper Sonora, spring view of, *140*
Upset Rapid, *189*
Upwarp, North American, 31–34, *33*
Urey, Harold C., 114–15
U.S . Geological Survey, 189–93

V

Valhalla Plateau, *116*
Vargas, Don Diego de, 165, 179
Vening Meinesz, Felix Andries, and convection current theory, *22–23*
Vermilion Cliffs, 44
Virgin River, *38,* 44–45, *91*
Vishnu Group, 67–75
Vishnu Schists, *28,* 68, 75, 95, 111
Vishnu Temple, *11, 83*
Vulcanism and volcanoes *(see also* specific features), 97–111, *100;* anatomy of a volcano, 98–99, *100,* 102; atmosphere, evolution of, and, 54, *54,* 55; distribution, *23,* 24, 98; faulting and, 98, 99; on Io, 97; Krakatoa eruption, 99; lava dams, *102,* 106–10, *107;* lava flow direction, 40; of Moon, 51; organic synthesis and, 115; primeval, 51–55, *54;* shield volcanoes, 98, 99; and Sinagua Indian settlement near San Francisco Peaks, 162; strato-volcanoes, 99; tallest volcano, Mauna Loa, 98; in Tertiary Period, 100–2, types of volcanoes, 99
Vulcan's Anvil (Forge), *103,* 110
Vulcan's Throne, 98, *103,* 106, *107,* 110; views from, *166, 169,* 193

W

Walpi Pueblo, 163, *163*
Wasatch Formation, *39,* 48–49, 79, 136
Wasatch Limestone, 48
Washburn, Sherwood L., 153
Watahomogi, 94–95
Watchman, *31*
Weeping Rock, 45, *45*
Wegener, Alfred, 13, 24
Wells, John W., 135
Wescogami, 94, 95
West Temple, 193
White Cliffs, 44
Whitmore Wash, 107, 110
Wilson, Edward Adrian, 12–13, 24, 25
Wilson Butte Cave, Idaho, 153
Wind River, 34
Wupatki, *151,* 162

X

Xanthus, 83, 85
Xenophanes, 83, 85

Y

Year, length of, 135
Yaki Point, *73*
Yavapai Museum, and view from, *177*
Young, G. M., 69

Z

Zion Canyon, 25, 31, *37, 38,* 44, 193; formation of, 44–48; fossils of, 87, 136, 142; Indians in, 158; life forms in, 145, 147; rocks of, 44, 45–48, *45, 57,* 59, *59,* 123; unconformities of, 65–67, 78–79, 136; views of, *31,* 45, *129;* vulcanism in, 99, *100;* youngest formation in, 48
Zion Canyon National Park, 189
Zoroaster Temple, *73, 83*